Highways in the Sky
Adventures of a Working Pilot

Highways in the Sky
Adventures of a Working Pilot

David B. Freeman

Nissi Publishing

Roanoke, Texas

Published in the United States of America by Nissi Publishing, Roanoke, Texas.

http://www.nissipub.com

ISBN 0-944372-13-9

Acknowledgements

No pilot flies alone, even when solo. My father introduced me to flying when I was a small boy. For this I am eternally grateful. Though I've had many flight instructors, a few stand out. Among them are CW2 Joe Phillips who first taught me to hover and fly a helicopter and CW3 Jim Miles who carried the burden of making me a helicopter instructor pilot. Pilots and crewmembers of the 57th Medical Detachment, both in Vietnam and at Fort Bragg all hold a special place in my heart and in my memories. J.D. Huss coached me through becoming a Certified Flight Instructor in the civilian world and in the process instilled in me the quest for excellence in all of my flying. Jeff Meaders, who ran the air service in my hometown for many years, trusted me with his airplanes and his students and taught me that courtesy and respect are as important to a pilot as his flying skills. Jim Gibson, Bob Duncan and Dennis Smeltzer are men with whom I've shared the cockpit many times, resulting in friendships that have lasted a lifetime. Jim Halek supplied the planes and the confidence that brought me back into aviation after many years of being groundbound. My wife Joyce not only lets me pursue my passions of flying and writing, but encourages me in both.

About the Cover

The cover was composed by the author using scanned images assembled in Adobe PhotoShop. The background is a portion of a 1968 Chattanooga Sectional Aeronautical Chart. The airplane is N98764, a 1976 Cessna 402B that the author flew professionally in 1977 and 1978. N98764 was one of Cessna's brochure airplanes that year and the image was scanned from a Cessna advertising brochure. The clouds and lettering were added using PhotoShop effects.

Table of Contents

Prologue

A License to Learn

Date: 9/17/70
A/C Type: Cessna 150
Registration #: N684RA
Route of Flight: San Antonio International - Local

It was a short check ride and when it was over I didn't know if I had passed or failed. We did the basic stuff, a few takeoffs and landings, some steep turns and slow flight, then the stalls. After I'd done a couple, the examiner took over and flew the last few minutes of the ride himself. He demonstrated something he called "the falling leaf maneuver" and seemed concerned that I had never seen it before. Basically, he just stalled the airplane and kept it stalled while it fluttered around the sky like a falling leaf.

Back at the airport I was just starting to go through the shutdown checklist when the examiner climbed out of the Cessna 150 and walked off. I thought I heard him say something about needing a bathroom. I was afraid I had made him airsick, maybe worse.

When I got inside I didn't see him anywhere, but his briefcase was still at the table where we had done the oral part of the exam. I sat down to wait. I had paid my money. He was going to have to give me the results of my flight test, one way or the other.

A few minutes went by and I saw him come out of the restroom. But, instead of coming toward me, he headed down the hall the other way toward the vending machines. Had he forgotten all about me? It was one of the greatest days of my life, but it meant nothing to him— just another day's work.

Finally he came back to the briefing table with a cup of coffee in his hand. He sat down across from me, fumbled around in his briefcase for a minute or two and pulled out a pad of forms. I was relieved to see they weren't pink. I watched eagerly as he wrote out the temporary certificate that qualified me as an airplane pilot. When he handed it to me, he looked me straight in the eye and cautioned, "You know this is just a license to learn, don't you?"

What an understatement! Passing the written, oral and flight examinations for my civilian pilot's certificate left me with a tremendous sense of accomplishment. It had been a hard journey, but now I was an airplane pilot! Little did I know how much there still was to learn.

As time progressed and I amassed ratings and certificates, actual instrument experience, and lots of cross country time, I began to realize just how much I didn't know in those early days.

A few years later when I became a flight instructor and had the opportunity to practice emergency procedures on a regular basis, I became even more thankful that I had never had a real emergency during my early hours of flying. I may not have been up to the challenge of a serious emergency.

Obtaining a license allows you to *be* a pilot, but actually *becoming* a pilot requires that you repeatedly do the things that a pilot does. You must use the airplane to go places—on sunny days and on cloudy days; when it's hot and when it's cold; at night; when it's raining, snowing or foggy. In the process, you learn the things that are essential to know if you are to defy gravity and nature on an ongoing basis and live to enjoy it.

The airlines have a built-in mentoring program that allows its pilots to amass experience while under the watchful eye of a seasoned veteran. Military pilots and business pilots have training opportunities that are not available to others who share the same sky. Regardless of our educational and training opportunities, we all learn through our own experiences and through sharing the experiences of others.

That's what this book is all about—shared experiences. Through a series of stories I invite the reader to share the cockpit with me on a variety of flights. Some are hardcore learning experiences, some are minor living experiences, and some are pure entertainment—awesome discoveries of the beauty around us.

The cockpit is a challenging classroom. At times it is boring. Occasionally it is a frightening place. Other times it is a place of wonder and adventure. It can also be a very lonely place and a place where you not only develop skills, but have the opportunity to develop character, insight, and understanding.

I am fortunate to have enjoyed more than 30 years of flying, in all sorts of aircraft and on a variety of missions: civilian and military, airplanes and helicopters, single-engine and multi-engine, business and pleasure. The episodes shared here occurred at varying levels of experience, resulting in insights that are applicable regardless of where you are in following your flying dream.

Each flight stands alone—kind of like the "I Learned About Flying From That" series in *Flying* magazine. I struggled for a while with how to group the stories. Initially, I wanted to group them into categories such as wind, instruments, or emergencies, but found some flights difficult to categorize and others that crossed boundaries into several categories. In the end I opted for a chronological structure. We pilots like logical structure anyway so it fits well within my ability to process information. Interspersed with the stories are some short blurbs to orient the reader with the progress of my flying career.

Happy Flying!

Chapter 1 – Beginnings

When getting to know a fellow aviator, I always ask about their early flying days. How did they get started? How and where did they accumulate time? What types of aircraft have they flown? What motivated them to become a pilot? These are the things that help you get to know a pilot. They tell you a lot about his or her desires and motivations, as well as the opportunities they were afforded and what they made of them.

In this section you will find a mixture of stories about civilian and military flying during my earliest flying days, beginning with my first fascination with flight and progressing through initial flight training. If you are a pilot you may identify with some of the experiences and emotions involved in these flights. If you are dreaming of becoming a pilot I hope you'll find the encouragement to continue.

A Young Boy's Dream

People learn to fly for a variety of reasons. Some justify it for business or economic reasons. Others admit it is pure ego, a love of machines—perhaps just a desire to be different. I fall into the latter category. The desire to be a pilot was a dream for me long before it became a reality.

The dream began on a lazy summer afternoon when the sun had set and the surface of the water was as smooth as glass. It was 1951 and I was three years old. Standing on the shore of Lake George in Florida I held onto my mother's hand and waited. Bass and bluegills nibbled at bugs on the surface of the water. Crickets and cicadas droned their symphony, welcoming the approaching twilight. Frogs croaked. I scratched, rubbed the sweat out of my eyes, and swatted at gnats. Over the creature sounds, another sound caught my attention. I looked up at my mother, who smiled, nodded, and pointed. Over the trees at the far end of the lake a small speck appeared. I watched as it grew in size until I recognized it as an airplane—a yellow airplane with huge floats under it. The plane dipped below the tree line and I almost lost it in the shadows. Then it appeared again moments later as it smoothly touched down on the surface of the water. Soon it was coming toward us, its floats creating trailing water plumes that looked like a pair of silver rooster tails.

As the Piper Super Cub approached the nearby dock, the tick-a-tick-a-tick of its idling engine matched the pounding of my heart. Mom led me out onto the dock to await its arrival. The engine was shut off and the propeller came to a stop. An arm reached out of the airplane's open window and threw a loop of rope around one of the dock's pilings. The airplane was pulled close to the dock and two men climbed out of it. One of them was my father. He lifted me in his arms and set me on the rear seat of the airplane. I took the control stick in my hands and imagined myself as the pilot. That's when the dream began in my young boy's heart.

Fast-forward seven years to a Saturday morning in the Spring of 1958 near Oxford, Mississippi. This time the airplane was a red and white Piper Tri-Pacer and it was my first time to fly. Ten years old and sitting in the front seat I was on top of the world.

The mission that day was to count fishing boats on the four big reservoirs in North Mississippi, a Game and Fish Commission project. I was able to go along for the ride because my dad was the boss and Harry Barkley, the pilot, worked for him. Dad sat in the back seat, allowing me to experience my first flight from the front seat. In front of me was a real set of pilot's controls. Never in my wildest imagination did I think I would actually get to touch those controls, but later in the flight Harry let me fly the plane as we flew straight and level between the reservoirs. He coached me as we flew along with comments like "nose too high" and I would push the control yoke forward only to start pulling on it again. It seemed logical to me that you had to hold the airplane in the air by pulling back on the wheel, but I soon learned that wasn't the case.

The flight was quite enjoyable until Dad spotted a flock of white birds on the backwaters of Sardis Reservoir. He called them "Mexican Wood Ibis" or something like that, and he and Harry began discussing how unusual it was to see them in that part of the country. Harry pulled up a lever that was on the floor between the seats which lowered the wing flaps. He began turning the plane in a slow circle so they could look at the birds. The plane turned, and so did my stomach.

Dad and Harry were so engrossed in the birds, they didn't notice my lack of enthusiasm. I was about to throw up all over the instrument panel. Dad caught on when he asked me a couple of times if I saw the birds and I didn't answer. He reached forward and opened the window on my side of the airplane and Harry leveled the wings. The fresh air blowing in my face did the trick, but I wasn't interested in looking at any more birds or counting any more fishing boats.

Harry distracted me by letting me fly some more and somehow they got their mission accomplished. More than that had been accomplished by the end of the day. My flying career had graduated from that of a three-year-old pretending to fly in the back seat of a Piper Cub on floats to that of a ten-year-old who had actually manipulated the controls of a real airplane in flight.

Eleven years later I took my first flying lesson. The in-between years contributed little to the flying dream except for one short thirty-minute ride on a DC-3 from Memphis to Oxford, and the fact that I read every issue of *Flying* and *Air Progress* Magazines in the school library from cover to cover. I also read the monthly aviation columns in *Popular Science* and *Popular Mechanics*, and all of the aviation books I could find by people like Ernest K. Gann, Richard Serling, Richard Bach, and Martin Caiden.

First Flying Lesson

Date: 10/22/69
A/C Type: Cessna 150
Registration #: N6564F
Route of Flight: Oktibbeha – Local

Little airplanes buzzed around the Oktibbeha County airport like flies around a pile of fresh manure. I always watched them when driving by, until one Sunday afternoon curiosity got the better of me and I turned in for a closer look.

The airport was hidden behind a grove of trees, so it wasn't visible from the highway. I followed a gravel road through the trees and pulled into the airport's gravel parking lot. Leaning against the car, I had an excellent view of the airport traffic pattern. I had no intention of going inside.

I have heard stories about people going to airports and being ignored by the people there. Prospective flight students say that when they go to an airport, they can't get anybody to talk to them. It didn't happen that way with me. Two college-age flight instructors noticed me in the parking lot and descended on me like car salesmen in a new-car showroom. Twenty minutes later one of them had me in the air in a Cessna 150. That alone would have been enough to have me added to the flight school's list of paying customers, but there was more. Under his supervision, I did the preflight inspection of the airplane, I taxied it to the runway (what an adventure that was), and I actually made the takeoff!

We spent forty minutes in the air that day. As we flew around in the vicinity of the airport, the instructor pointed out landmarks, explained how the controls in the airplane worked, demonstrated how to make turns, climbs and descents, then gave me the controls so I could imitate what he had done. Fortunately, the little Cessna was forgiving and obeyed my tentative control inputs. At least half of the time we were in the air, I was actually doing the flying. All too soon the airplane's nose was pointed toward the runway and the introductory flight was over. I had no problem booking a second lesson for the following week.

Financially, it was tough. I was working as a horse trainer, making about a hundred dollars a week and making payments on a new car. I started spending $25 to $40 each week on flying lessons. To me it was worth it. I looked forward to every lesson with eager anticipation.

Sounds of Silence - Number One

Date: 10/19/69
A/C Type: Cessna 150
Registration #: N1825J
Route of Flight: Oktibbeha – Local

It was my seventh lesson. My instructor for the day, Doug Hardy, discussed forced landings with me before the flight, so I thought I would be ready. I was concentrating on my flying, but at the same time I was watching Doug out of the corner of my eye, knowing that at some point in the flight he would retard the throttle and announce "forced landing" because that's what he told me he would do during the preflight briefing. I would then be expected to execute the appropriate emergency procedures, select a landing spot and maneuver the airplane into position to land there.

But it didn't happen that way. We were climbing toward the practice area when the engine quit. One second it was running, the next it was deathly quiet. Doug had not touched the throttle. This was the real thing.

I waited for Doug to do something, but he didn't. I was in the left seat, so it was up to me. Still, having Doug there gave me confidence and I knew what to do because we had rehearsed the engine-out drill over and over before the flight and on previous flights. Since Doug was making no effort to do anything about our plight, I started doing what I had been taught. I lowered the nose of the airplane to establish a glide. I checked to make sure the mixture was pushed all the way in to the full rich position and that the magneto switches were on. Then I checked the fuel selector switch. It was off! How could it be off? I didn't know, but quickly moved it to the "on" position and the engine caught and sputtered back to life.

It had been a risky thing to do, but Doug had turned the fuel selector switch off while I was looking outside of the plane for other traffic. It wasn't hard for him to do that without me noticing since the fuel selector switch was on the floor between the two seats, but within easy reach from either seat. The engine had run on a minute or two, until the fuel had burned out of the lines, so when the engine quit, Doug was sitting innocently with his arms folded. He must have had a lot of confidence in his own ability to land somewhere other than the airport in case the engine didn't restart, but I'll tell you it got my attention. I was pleased that I did the right thing, but Doug wasn't through. As soon as I had gotten the engine running and started climbing back up to altitude, he pulled the throttle back to idle and announced "forced landing." Now I had to pick a field and try to get in it. I failed miserably at the first attempt, and the second, and the third. By the fourth try, I was starting to figure out that those ground track maneuvers we had practiced earlier in our training had something to do with maneuvering the airplane into a good landing position, with or without the engine running.

When trying to land safely after loss of engine power, there is a lot to factor in. In addition to establishing a glide at the proper speed and trying to find and correct the reason for the power loss, you have to find a suitable landing spot, then maneuver the airplane into position so you can land in the spot you have chosen. The wind speed and direction and your altitude all affect how you must maneuver the airplane. The objective of the lesson was to prepare me to handle an engine failure in case one occurred when I was flying solo. It took repeated practice to get it right.

Blowing in the Wind

"The answer my friend, is blowing in the wind."
Bob Dylan

December 1969 – A day in which I didn't fly

The storms ended on the Tuesday night before my next scheduled flying lesson, giving birth to a Wednesday full of sunshine and blue sky. I went about my morning chores with anticipation. We were going to work on takeoffs and landings that day. I would solo soon; confidence was high.

I worked the horses that belonged to the paying customers, but while riding them my mind was far away in the sky. There were a few other things I had to take care of around the stables and I hurried through those tasks. Finally, noon approached and most of the morning chores were done. My lesson was scheduled for two o'clock and I wanted to hit the road early to allow plenty of time to get to the airport, go over my lesson plan, and preflight the airplane before the time I was scheduled to meet with my instructor.

Highway 45 was straight as an arrow, but had its share of ups and downs due to the rolling terrain of the countryside and the age of the concrete. My Malibu Super Sport liked the open road, but I knew there would be a highway patrolman somewhere between Macon and the Highway 82 intersection. I didn't want anything to make me late for my flying lesson so I kept the Malibu's speed just under eighty miles per hour. With all four windows rolled down, the crisp, cool air made the ride invigorating.

I spotted the state trooper in the parking lot of the Colt 45, a popular honky-tonk just south of the Highway 82 crossroads, in time to roll past him at a decent sixty-five. A couple of minutes later I turned off the highway and drove up the gravel road to Oktibbeha Airport.

Approaching the airport, I expected to see the sky filled with airplanes as usual, but none were to be seen. With an active ROTC program at nearby Mississippi State University, the traffic pattern at Oktibbeha was always a busy place. Camp's Flying Service had six Cessna 150 trainers, most of which were in the air at any given time. Today there were no planes in the traffic pattern. I was puzzled, but not overly concerned. Maybe they were all out at the practice area. That was fine with me, for my entire lesson that day was to be made up of practicing short and soft field takeoffs and landings, and if the wind was right, crosswind landings. Less traffic in the pattern would make it easier.

I rounded the last curve and discovered an empty parking lot except for two or three cars near the office entrance. That was so unusual I wondered what national emergency had occurred.

Doug Hall, my flight instructor, was standing at the dispatch counter when I went inside. "There you are," he said. "I tried to call you before you left home to tell you not to come."

"Why? What's up?" I asked.

"The wind," Doug replied. "It's much too windy to fly today."

The wind? I knew that clouds would have kept us from flying. Rain or nighttime would have kept us on the ground at my experience level, but the wind? It simply hadn't occurred to me that the wind would ground an airplane. The wind was an airplane's element.

Doug chuckled at the puzzled look on my face. "Come on," he said, "Let's go outside." I followed him through the door to the flight line and there stood all six of the Cessna 150's, securely tied down with their wings rocking in the wind.

"Look at the windsock," Doug said, pointing to the white windsock that hung on a pole by the maintenance hangar. It was standing straight out, horizontal to the ground. Even the pole was slightly bent because of the wind. "Imagine trying to practice landings in that," he said.

"Wow," I replied. "I didn't even think about that. It was such a pretty day I couldn't wait to go flying."

"That's okay," Doug said. "But, that's why I tried to call you before you left home. Now that you're here, let's go inside and have a ground school session. That will give me a chance to go over crosswind landing techniques with you again."

What followed was an explanation of crabs and slips, using a model airplane to describe the maneuvers. It all made sense, and the maneuvers should have been simple to perform. But I knew from experience they weren't.

The wind at Oktibbeha normally blew right down one of the runways. Consequently, I had not experienced many problems with the wind during my initial stages of training. But now that I was getting ready to solo, Doug wanted to make sure I could handle takeoffs and landings with the wind blowing across the runway instead of straight down it. To accomplish this we would either use a runway other than the active at Oktibbeha, or we could go over to West Point and use their runway. I had to master crosswind landings before I could solo. You never knew when the wind might shift directions while you were in the air. But not today. The wind was just too strong.

Doug and I spent an hour together, and afterwards I drove home to groom and feed the horses. Later after the sky had grown dark I sat on the steps outside my mobile home and looked up to see millions of stars sprinkled against a totally black sky. There were no clouds and the moon had not yet risen.

Even now, long after the sun had set, the wind continued to howl through the trees and rattle the tin on the barnyard roof. The animals were quiet, leaving the sound of the howling wind to permeate the night. I wondered why I had paid so little attention to it before. As a pilot I would have to be much more conscious of, and in tune with, the elements. They would be my adversary or my ally, depending upon their nature at the time and my ability to work with, rather than against, them.

With the night came an autumn chill that caused me to shiver and I soon headed for the warmth of the house. I heated up some soup and made a grilled cheese sandwich. As I sat down to eat, I opened my flight instruction manual to the section on weather and began to read about cold fronts and the effects of their passage.

First Supervised Solo

Date: 1/15/70
A/C Type: Cessna 150
Registration #: N2123X
Route of Flight: University-Oxford – Local

Circumstances changed with my job and I moved back to Oxford, Mississippi, my hometown. I started working at a clothing store and again had a little money for flying lessons. The flight school at Oxford had Piper Cherokees instead of Cessna 150s. I began flying again with Jerry Cockrell as the instructor. We flew up north of town and Jerry gave me a forced landing near the Tallahatchie River Bridge. There were woods all around, but on the west side of the road near the highway was a field. I headed for the field but didn't slow the airplane to the proper speed and the Cherokee floated right over the only reasonably safe landing area in sight.

Back at the airport I had trouble landing the Cherokee because it was a low wing airplane and it tended to float when flaring for landings. I just wasn't getting it right. Jerry knew that I was anxious to solo, and that most of my previous time had been in Cessna 150s, so he suggested that we try our next lesson in the one Cessna 150 that the flying school had. It was an earlier model than the ones I had been flying and had a straight tail instead of the swept tail on the newer 150s. Jerry scheduled me for another lesson late that same afternoon.

The winds late in the afternoon were almost totally calm and I was more comfortable flying the Cessna 150. I made decent landings in the 150, but not as good as I thought they should be. After we had done a few, Jerry told me to taxi back to the ramp. I figured he'd had enough of my flying for one day. As soon as I turned off the runway and onto the taxiway, he had me stop the airplane and he asked me for my student pilot's certificate. I honestly thought he was going to take it away from me and tell me I should forget flying altogether. Instead, he endorsed it for solo flight, handed it back to me and got out of the airplane. "Give me three takeoffs and landings," he said. Then he cautioned me that the Cessna 150 would perform differently with only one person on board. He said it would climb faster and that the turn to downwind would seem faster. He closed the door and I was on my own.

It is hard to describe the feeling that you have when first being allowed to take an airplane up by yourself. You feel both proud and nervous at the same time. But there is so much to do that you don't have time to dwell on being nervous for long. You must remember to do all of the radio calls yourself, look for traffic, monitor airspeed, altitude and keep up with where you are all the time. In addition, you must constantly be on the lookout for other traffic. If the instructor has done his or her job properly, you are ready, even though you may not feel that you are.

When I turned downwind the first time it felt weird as the little plane accelerated more rapidly than I was accustomed to. Not only that, but the plane climbed to pattern altitude more rapidly than what I had previously experienced. I was glad Jerry had warned me about how the airplane would behave differently with only one person on board. On landings the plane didn't settle as quickly, but it was still the same type of airplane that I'd been flying for about 10 hours. The three takeoffs and landings were over before I knew it and I was taxiing back to the ramp, one proud young student pilot.

Jerry met me at the tie-down area and congratulated me. When we got inside, he reached into a desk drawer and pulled out a big pair of scissors. He told me to pull out my shirttail. Then he turned me around and cut a huge chunk out of the shirttail in the back. Using a black magic marker, he wrote my name and the date on the shirttail and stapled it to the wall in flight operations. This was a tradition in those days that seems to have been lost in recent years. I guess that shirttail was thrown away long ago, or is in some storage box somewhere. I wish I had it now, but I didn't think of that for a long time after it was no longer around.

The shirttail was significant for another reason. Without realizing it, I was wearing one of my dad's shirts that day and somehow, after it had been cut, it got washed, ironed and hung in his closet. I was in his bedroom the day he pulled it out to put it on and noticed the shirttail had been cut. There was a big "u-shaped" cut right in the middle of the back. "What is this?" he wanted to know. That is when I told him for the first time that I was learning to fly and that I had soloed. I honestly didn't know how he would feel about it. To my surprise, he was pleased. That was a relief. I hadn't actually been sneaking around about learning to fly, but I wasn't advertising the fact, either, since I didn't know how my parents would react.

Warrant Officer Flight Training

My first solo flight occurred in January, 1970. A few weeks later I learned that I was about to be drafted. This was during the Vietnam war and I had already decided that if I had to participate in the war, I would like to do it as a helicopter pilot. One of the benefits of living in a small town is that everybody knows everybody else, so when my draft notice arrived at the local draft board, the secretary there called my mother and told her that if I didn't want to be drafted into the Army the next day, I had better go enlist that day.

I visited the Navy recruiter and the Air Force recruiter and learned that I had to have a college degree to get into flight school in either of these services. The Army recruiter didn't care that I had only completed three years of college. He had a quota of potential helicopter pilots to sign up and I would do fine, thank you very much. So began my quest to become a US Army helicopter pilot.

I went to basic training at Fort Polk, Louisiana, in July. After that I was sent to Fort Wolters, Texas, to begin flight training in the US Army Rotary Wing Warrant Officers Course. At Fort Wolters, we were WOCs—Warrant Officer Candidates, the lowest scum of the earth. We spent six weeks in preflight classroom training before starting to fly. When we did start flying it was November, with cold and windy days.

Hovering

November 1970

Flying a helicopter in the air is not hard. You just point the main control stick, called the *cyclic control* in the direction you want to go and the helicopter will pretty much go there. You can fly forward, sideways or backwards. It is nimble and responsive and fun, especially if it is moving forward and the aerodynamic forces of the forward motion are helping the pilot keep things streamlined.

Put a helicopter close to, but not on the ground, however, and it's a whole different story. The instructor pilots (IPs) wanted us to pick the helicopter up from the ground to a three-foot hover and keep it stationary over the ground. Right! The problem with that little task is that too many things happen at once.

The thing that makes the helicopter go up is called a rotor. Depending on what type of helicopter you are flying, it has two, three, four, perhaps even five rotor blades attached to a hub and spinning around at an incredible speed. To go up or down, an angle, called *pitch* is applied to the rotor blades. Increasing the pitch angle of the rotor blades causes the helicopter to go up. Decreasing the pitch angle causes the helicopter to go down. The pilot does this with a lever called the *collective*, which is mounted to the floor at a forward sweeping angle just to the left of the pilot's seat. It is called *collective* because when you pull it up, the same amount of pitch is applied to each of the rotor blades *collectively*.

There's another control that manipulates the rotor blades. This one is called the *cyclic* and in most helicopters it is directly in front of the pilot, sticking up from the floor. In some modern helicopters, the cyclic hangs from the ceiling. The purpose of this control is to tilt the rotor system in the direction you want the helicopter to go. Moving the cyclic control tilts a swashplate on the rotor system, which then causes different angles to be applied to the rotor blades at different points in their rotation, effectively tilting it forward, sideways, or backwards, corresponding with the way the cyclic control is moved. Push forward on the cyclic and the rotor system tilts forward, causing the helicopter to move forward. Push sideways on the cyclic and the rotor system tilts sideways and the helicopter moves to the side.

Got it, so far? Good, but there's more. Because the rotor system is rotating, it causes the body of the helicopter to try to rotate in the opposite direction. Remember the laws of physics? For every action there is an equal and opposite reaction. To keep the body (or fuselage in aviation terms) of the helicopter from rotating, there is a tail rotor. This tail rotor has pitch, too, which is controlled by a pair of pedals. Pushing the left pedal causes the tail of the helicopter to go to the right and the nose to go to the left. Pushing the right pedal causes the tail of the helicopter to go to the left and the nose to the right.

Now you've got three controls to work with, none of which are aided by friction when you're in the air. That means they need constant input from the pilot. There's another control we have to add into the picture—the throttle. When you load the rotor system up by increasing the pitch angle in the blades, it causes the engine to want to slow down, but you can't allow that, because if the engine slows down, the rotor system slows down and if the rotor system slows down, it can't sustain lift and the helicopter will fall to the ground. So, when you increase collective pitch, you must at the same time increase engine power with the throttle, which is like the twist-grip throttle of a motorcycle. The throttle is at the end of the collective pitch lever. When you're pulling the pitch lever up, you must at the same time be rolling the throttle clockwise to add more power and when you're lowering the collective pitch lever, you must at the same time be rolling the throttle counter-clockwise to decrease engine power so you won't get an engine or rotor overspeed condition.

In time, all of this becomes second nature, but when you're first learning to manipulate all of those controls, it is overwhelming. The IPs know this, so they let you handle one control at a time. Basically, they get you all by yourself in a field about the size of a football field and they let you handle the cyclic until you can maintain a position reasonably well over the ground. After that they let you take the collective. Then they slowly add the pedals, then the throttle. They've got guts, these IPs that teach new students how to hover, because when you're learning, you're all over that field. Not just sideways, either. You go up and you go down, you go around and around, and you tilt forward and you tilt backwards and the tail swaps place with the nose and you sweat and the IP laughs, then he takes the controls and the helicopter settles down into a perfectly smooth hover, then he gives you first the cyclic, then the collective, then the pedals, then the throttle, and it starts all over again. It isn't easy.

I can attest to the fact, however, that once you learn it, you don't forget it. I climbed into the front seat of a Huey recently after not being in one for eighteen years and picked it up to a smooth, steady hover. It felt like the last time I had been in one had been just a day or two earlier.

Learning to Talk
November 1970

Today I can fly into New York City on an instrument flight plan and not be intimidated by the instructions given by the air traffic controllers. But that's a learned skill. Sometimes I take it for granted, but even with years of experience it takes concentration to listen to the radio calls that apply to your aircraft and to communicate clearly and effectively with air traffic control. I have some notes from early in my helicopter training about how we learned to use the radio in our pre-solo days. The notes remind me of the steps we often have to take to learn the basics.

Learning to use the radio is an important part of every student pilot's training, but in a helicopter there are some added challenges. Because helicopters are basically unstable, it takes both hands and both feet to fly one. At least it did back when we had to start using the radios. The instructors took care of all of the radio calls during the earliest part of our training. When it came time for us to start handling that chore, you'd have thought we were a bunch of two-year-olds.

The first problem was just talking. Even when you thought ahead about what you wanted to say, as soon as you keyed the microphone, your brain would go dead and your tongue would become about four times thicker than normal and you'd trip all over you words. The basics of air traffic control communication are the same basics that apply to most conversations: who, what, when, where; sometimes how and why. Imagine you're a newly-soloed student pilot taking off from the main heliport at Fort Wolters and when you're flying away from the area, the tower controller says, "Frequency change approved."

Problem one is what frequency do you change to? Problem two is which of your totally occupied hands do you divert to the radio to turn knobs? Once you find the next frequency and get it tuned, what do you say? Fortunately, our instructors knew this would be a formidable challenge for us in the early days, so they taught us to break this seemingly impossible task down into several smaller tasks, each of which we should be able to manage. It went something like this:

1. Before takeoff, look up all of the frequencies you might need during the flight and write them on the inside of the aircraft windshield with a grease pencil. Make sure to write them where you can easily see them but where they don't block your view.

2. When it becomes necessary to change frequencies, glance at the radio to see what frequency it is currently on.

3. Mentally calculate the number of digits and the direction necessary to turn each of the frequency control knobs to arrive at the correct new frequency. (On most radios it takes two knobs to make the frequency change, one for the digits left of the decimal point in the frequency and the other for the digits to the right of the decimal point.)

4. With your left hand, tighten the friction on the collective pitch lever so that the lever will not move up or down on its own accord.

5. Tighten the throttle friction to maintain RPM.

6. Transfer your left hand to the cyclic control and your right hand to the first radio frequency control knob.

7. Turn the knob in the predetermined direction and number of clicks to the new frequency. While you're doing this, keep up your instrument scan and your constant vigil for traffic.

8. Quickly glance at the radio to make sure you correctly selected the new frequency. If not, make the necessary adjustments.

9. Now do the same thing with the other control knob.

10. Place your call to the air traffic controller on the next frequency.

I know. You're reading this and it sounds like overkill. But remember that learning to fly a helicopter is like learning to roller skate on a beach ball on the ocean in a storm. This is what it took at first, but by our second week of solo flying, we could skip most of the steps and just tune and talk.

First Supervised Solo - Helicopter

Date: 11/17/70
A/C Type: TH-55
Registration #: 16446
Route of Flight: My Tho Stage Field, Fort Wolters, Texas

Seventeen hours in the TH-55 helicopter and I still hadn't soloed. The hovering part was hard enough, but now I couldn't seem to handle any wind above 15 knots.

Usually I slept on the bus ride to Downey Army Heliport, but not this time. This time I worried. My IP, CW2 Phillips was pulling for me, but his hands were tied. He and I both knew that at this point in my training the Army was looking for ways to wash out Warrant Officer Candidates that weren't going to make the grade.

Many of the guys in my flight class had soloed after eight to ten hours. Twelve was considered the norm, and fifteen was supposed to be tops. I was over the limit, but Mr. Phillips said he had confidence in me. He knew I was going to make it, but just to keep him and me both out of trouble, it had to be today. You'd have thought that since I had already soloed in airplanes, this would be easy. That's what I thought, but it wasn't the case.

Wouldn't you know it would be windy again? Why did they have to train us on the West Texas prairie where there were no trees to block the west wind. It hit us in all of its fury, a biting cold wind that chilled us to the bone and tossed our Mattel Messerschmidts around the sky like little scraps of paper. Why couldn't I have been scheduled for the morning flight? At least then I might have a chance to solo before the wind had a chance to build up to its afternoon proportions.

After the briefing at Downey Heliport it was time to ride the bus again. My stick buddy, John Almeda flew the first period. Mr. Phillips rode with him in the helicopter to My Tho stage field, where John would practice takeoffs and landings in the traffic pattern for the rest of his flight training period. John had soloed three days earlier.

After the bus ride, I sat in the shack at My Tho and looked out the window at the TH-55s in the traffic pattern. It was too cold for me out there. It was cold inside, too, but at least there was no wind. Mr. Phillips stood out by the flight line with the other IPs who had students flying solo. He didn't look cold. I was shivering. Maybe it was from the cold, maybe it was nerves.

Looking around the room, I was very much aware of the fact that, except for Jim Johnston, I was the only one in the flight who hadn't soloed. Jim was not going to make it. He knew it and so did everybody else. I wondered if they thought that about me, too.

After an hour and a half, Almeda set the TH-55 down in the refueling spot. Now it was my turn. John had a little trouble with the wind, bouncing a few times before lowering pitch, but then the helicopter was down. Mr. Phillips walked out to meet him. He shook John's hand as John climbed out of the helicopter. John was in like flint.

I wished I could get rid of the nervous feeling in the pit of my stomach. I prayed for God to take away the fear and give me peace, but I didn't sense an immediate answer.

The helicopter was refueled, and I went out to do my preflight. Then, I got in, started the engine and engaged the rotor blades. Mr. Phillips came out, climbed in and plugged his flight helmet into the intercom. "Okay, Freeman, today's the day. Give me three good trips around the pattern and it's all yours."

"Yes, sir." I looked around to make sure we were clear, then eased up on the collective. Remembering what he had drilled into me, I got the helicopter light on the skids and felt around for a neutral cyclic before breaking ground. I took a deep breath and let it out, then willed myself to breathe normally. That concept was quickly forgotten, as holding my breath, I lifted the little Hughes to a hover. In a flash it was all over the sky. The tail tried to swap ends with the nose as I let it drift sideways toward the middle of the field. I knew I must stop it, knew I must settle down and keep it steady, but my hands and feet didn't get the message.

I fought it. The controls were all over the cockpit. Mr. Phillips didn't grab for the controls, he just talked to me. "Take it easy, Freeman. Stop fighting it. You're making your own problems over-controlling. Just settle down."

That was easy for him to say. I honestly believed he had a "hover button" on his side of the cockpit. Whenever he took the controls, the aircraft instantly assumed a rock steady hover. But he wasn't taking them this time. He was going to let me kill us both.

"Rest your arm on your leg and hold the cyclic with the tips of your fingers," he coached. "Keep your feet still. You're letting this little helicopter get the best of you."

It is better than me, I thought, then instantly banished that thought. I could do this, I *must* do this. I allowed the tension in my body to relax. I rested my arm on my leg and held the cyclic with my fingertips, just like he said. I managed to quit fighting the pedals and kept the nose pointed reasonably straight ahead. The helicopter stopped drifting. It was sinking and I somehow recognized that it was and eased up on the collective. *Easy*, I told myself, and I listened! The sink stopped and we didn't go shooting into the air. I even kept the RPM up. Finally, I was in control!

"Okay, we can leave the parking area," Mr. Phillips said. I had managed to get the helicopter off the ground and into a three-foot hover. Step one.

The takeoff was a breeze. So was flying the traffic pattern, though I made the turn to downwind a little too tight. When we were abeam the landing spot, I lowered the collective and started my turn. I began bleeding off airspeed. It was here that I got into trouble.

Three problems attacked at once: I bumped the throttle into the overspeed governor as I decreased collective pitch—not once, but three times. That meant I wasn't coordinating my throttle and collective movements properly. The gusty wind was buffeting the helicopter and I was having a hard time compensating. The closer to the ground we got, the more pronounced the effect seemed and the more I fought it. Then, there was the drift. A strong right quartering crosswind was blowing us left of course.

The three problems combined seemed more than I could handle. But at least I had identified and evaluated them. Did that mean I was developing what Mr. Phillips called "air sense"? I hoped so.

If I had identified the problems, maybe I could lick them. Time went into slow motion and I began working out the solutions. A murmur of approval came from the right seat. "Now you're getting it," Mr. Phillips said. I beamed with pride and almost lost it all. But, I was quick to recover. I was on top of it.

My subconscious began giving me instructions. *Back off a couple hundred RPM and grip the twist throttle with just your fingers, not the whole hand. Don't fight the buffets. Just concentrate on keeping the right power setting and the right descent angle, and leave the buffets alone. Crab into the wind. Look at the far end of the stage field strip for a better perspective.* I was remembering all the things Mr. Phillips had taught me.

We made it down, and then up again. We did it two more times and Mr. Phillips seemed pleased. "All right," he said. "Let me out. It's time you did this alone."

Hovering back out to the takeoff pad after dropping Mr. Phillips off, I was too busy to think about the fact that I was finally soloing. But when I made the first turn to the downwind leg it felt like somebody had just given the little TH-55 a swift kick in the behind. Mr. Phillips had warned me that would happen, and the same thing had happened when I had soloed in the Cessna 150. I was still caught off guard, and panicked, but just for an instant. By the time I was midway through the downwind leg, I had managed to settle down and talked myself through the rest of the flight. I completed the three takeoffs and landings, and turned off the landing pad to hover back to the parking area where I would pick Mr. Phillips up for the ride back to the main post. The takeoffs were a snap, and the landings were okay. It was the hovering back to the ramp and trying to park that nearly did me in. It was the wind again. It was blowing from every direction except from directly in front of me where I could handle it. The first time I tried to set the little Hughes down, the wind kicked the tail up, pitched me forward and I scooted across the ramp toward the helicopter in front of me. I responded with too much aft cyclic and suddenly I was moving backwards toward another helicopter that had just pulled in to park behind me.

I pulled the Hughes up to about ten feet, got things under control, and then started easing her back down. Again the wind whipped the tail around and again I fought it. I saw Mr. Phillips out of the corner of my eye, standing about twenty feet away, shaking his head. I had to show him I could do it. I took a deep breath, willed everything to *be still*, and then bottomed pitch. It wasn't pretty, but the helicopter was on the ground, reasonably close to the marked parking spot, and still in one piece.

Mr. Phillips ducked his head, went around the front of the helicopter and climbed in on the left side. "I don't know," he said as soon as he had put on his flight helmet and plugged it in. "You were doing all right until right there at the end. What happened?"

I started to reply that it was the wind that had been giving me fits, but that sounded so feeble, since the other students had been handling it, so I kept my mouth shut.

"I've got it," he said. "You sit back and relax for a while."

He picked that little helicopter up to the finest, smoothest three-foot hover you've ever seen. It made me sick. Then the wind hit the tail and we nearly swapped ends. Suddenly, he had his hands full trying to keep the little Hughes in one spot over the ground. I looked over to see a bead of perspiration on his forehead. So, it wasn't all me, after all. There really was a butt-kicking wind out there. I was grinning on the inside, but I sure didn't let Mr. Phillips see it.

On the way back to Downey Army Heliport, we had a conversation that encouraged me.

"You can't see the wind," Mr. Phillips started explaining, once we are airborne, "But you can see its effects." He was flying; I was riding and listening. This was one of those times with him I treasured. It was a time when he was really teaching me, not browbeating me for something I couldn't seem to do right.

"You watch the tops of trees, the way the water moves on a lake or pond, the way smoke curls, the direction a herd of cows faces, and you learn what this invisible wind is doing. Then you make your adjustments accordingly. Sometimes the wind is beyond the limits of the aircraft. When it is, you sit on the ground—if you can. Sometimes you can't. So you learn how to work with the wind, not against it. It will take you a lifetime of flying. The wind can always rise up and bite you. But make it your purpose to rule it, not let it rule you. Know what it's doing, try to anticipate what it's going to do, and always consider it in your plans."

I listened while he flew and talked, nodding my head as if I understood it all. The idea that the wind could always rise up and bite you, no matter how much experience you had, was not too comforting. I knew he was right and resolved to learn all the techniques I could for recognizing and dealing with the wind.

A Mixed Bag - Flying Airplanes Again

The Army Aviation Rotary Wing Course was nine months long. The first five months were done at Fort Wolters, Texas, and the final four months were done at Fort Rucker, Alabama. At Fort Wolters, we completed Primary I, which consisted of learning to fly the helicopter and Primary II, which consisted of confined area and pinnacle operations, navigation and night flying. All this was done in the Hughes TH-55, nicknamed the Mattel Messerschmitt because some of its parts were actually manufactured by the Mattel Toy Company.

The first flying we did at Fort Rucker was instrument training in the TH-13T, which in civilian life is called a Bell 47G. This is the helicopter you see used in Korea on "MASH" episodes. We didn't actually get instrument rated at Fort Rucker, but we supposedly flew enough instruments to get us out of trouble if we inadvertently flew into the clouds. The final stage of training at Fort Rucker was called Tactics and here we flew Hueys, the workhorse helicopter of the Vietnam War.

Upon graduation from flight school, we were Warrant Officer Aviators with approximately 210 hours of helicopter flying experience under our belts. We went en masse to the FAA office and took a written test on the FAA rules governing civilian flight. Passing this exam made us each an FAA certified commercial helicopter pilot.

I was sent to Fort Sam Houston for a couple of months of training for qualification as a medevac pilot prior to my all-expense paid trip to Southeast Asia. While in San Antonio, I decided to add a fixed-wing rating to my new commercial certificate. That's why I was flying one of the Fort Sam Houston Flying Club's Cessna 150s out of Martindale Army Airfield one afternoon when I got another lesson about the wind.

Crosswind Landing at Night

Date: 9/9/71
A/C Type: Cessna 150
Registration #: N3830J
Route of Flight: Martindale Army Airfield – New Braunfels

What a beautiful day it was! I preflighted the little 150 and took off to the east, where I spent a few minutes in the practice area doing ground track maneuvers. With just a slight breeze coming out of the south, my turns about a point and "S" turns across a road went well.

Satisfied that I had the ground track maneuvers nailed, I climbed to altitude and practiced slow flight and a series of "power on" and "power off" stalls and recoveries. *No problem there*, I thought and headed up to New Braunfels to practice short and soft field landings.

The landings went well, and I felt the satisfaction of having what my instructor, J.W. Chapman, would have called "a good lesson." On the way back to Martindale, I gave myself a practice "forced landing," shutting off the throttle, picking out a good emergency landing spot and gliding toward it. When I was satisfied I had the landing spot made, I added power and climbed back to my cruising altitude of two thousand feet. The check ride coming up in a few days was going to be a snap.

I noticed when following the highway back to Martindale that it took a pretty good crab angle just to stay on course. The sun was also setting. I had been having so much fun during one of my few times alone in an airplane that the time had gotten away from me. The wind had shifted to the northwest and had picked up considerably. I picked up the microphone and called the airport.

"Martindale Tower, Cessna three-eight-three-zero-juliet, five miles northeast for landing."

No response.

I tried again, "Martindale, Cessna three-zero-juliet."

Still no answer.

I checked my watch. It was after 6:00 p.m. and it was a Sunday afternoon. The tower had closed for the evening. I flew toward the traffic pattern keeping my eyes open for other traffic and announcing my position every so often on the tower frequency.

On downwind I noticed the windsock was standing straight out and the direction was about 45 degrees from the runway heading. Earlier, the winds had been light and variable out of the south, but now they were out of the northwest, and much stronger. Apparently, the cold front predicted to arrive later that evening had arrived earlier than forecast.

On final, it seemed to take forever to reach the runway. I had to keep adding power to get there. As I got lower, the wind started blowing the little plane around a lot and I found myself over-controlling. The closer I got to the ground, the harder it was to stay aligned with the runway.

The runway lights were on and sunset was fading into twilight as I struggled to keep the plane between the two rows of lights. The wings were rocking from side to side and the airplane's nose was pointed way off to the right. This would be a crosswind landing, and at night, too!

I didn't panic, but I knew the landing wouldn't be pretty. Thank God no one was there to watch me. Before I was ready for ground contact, before I had kicked in enough rudder to align the nose of the airplane with the runway, the left wheel touched down and there was an awful skidding sound. The plane slid across the runway to the left and it wasn't pretty. I thought the wing tip was going to scrape the ground. Then the nose of the airplane jerked around and I attempted to keep it lined up with the runway. I was supposed to do that *before* I let the Cessna touch down! My feet danced on the rudder pedals as the plane rocked from side to side, then came the awful sound of metal scraping on pavement. Finally, the plane slowed enough that I had it under control, but the ride felt awful lumpy. Something was wrong.

I stopped the Cessna, opened the door and looked at the wheel on my side. It looked all right. Then I leaned across the small cockpit and popped open the door on the other side. Uh-oh, trouble! The tire was flat. No wonder I couldn't taxi straight. My skidding on the runway had separated the tire from the rim. It was then that I began to appreciate the placard on the instrument panel—"Demonstrated Crosswind Component – 17 Knots."

Cross Country

Far, we've been traveling far, but not without a hope, and not without a star.

From "America" by Neil Diamond

Date: 10/4/71
A/C Type: Cessna 172
Registration #: N9898G
Route of Flight: Oxford – Little Rock – Memphis – Oxford

The cross country solo flight in the Army's Primary Helicopter Course had been a joke. Forty little orange helicopters followed one another from Fort Wolters to Fort Sill, Oklahoma, and back with a fuel stop in Bowie, Texas, on the way up. We didn't actually need the fuel, but in order for it to be a "real" cross country, we had to land at three different places.

The WOC flying the lead helicopter had been a flight instructor in civilian life and had logged over 3,500 hours as a pilot before he joined the Army. Two of our IPs kept watch on us from above in a fixed-wing Cessna T-41. We weren't going to get lost and our navigational skills were not challenged.

We did our stint in navigation class learning to draw course lines on sectional charts and marking checkpoints every ten miles or so. We learned to use the E6-B calculator to determine the heading we would need to fly to correct for drift based on the forecast winds. We did a little practice navigation, all from about 1500 feet above the ground. But when we flew our cross country, we basically just followed the helicopter in front of us.

When our class graduated from the US Army Rotary Wing Aviator Course at Fort Rucker, Alabama, we all took the FAA written examination for a commercial pilot's certificate with a helicopter rating. I went from Fort Rucker to Fort Sam Houston in San Antonio for the medevac pilot course. While there, I got a little more instruction in airplanes and took the check ride that added a single-engine airplane rating to my commercial certificate. That's how I became a commercial pilot with a little over 240 hours in my logbook and virtually no cross country flight experience.

The first time I got a glimmer that things in real life were a lot different than the basic line-on-the-chart, checkpoint-every-ten-miles procedures taught in flight school, I was a passenger.

The ink was dry on my helicopter ticket, but just barely dry on my airplane rating. In thirty days I would be in Vietnam, but for the time being I was spending a little time at home. With time on my hands and a girlfriend to impress, what else was there to do but go to the airport and rent an airplane?

Candee Vincent and I showed up at the Oxford airport and were greeted with a surprise—an opportunity to fly without it costing me anything. Hal Freeland, a local attorney and also a pilot, needed to go to Little Rock, but he would be staying there and the plane needed to come back to Oxford. But it also needed to go to Memphis and have one of its radios dropped off at the radio shop. It didn't take much arm twisting to get me to agree to the trip. Hal had no problem with Candee coming along.

Hal was the one paying for the plane, so he would be the pilot on the way to Little Rock. I didn't observe him doing any flight planning, but he did call Flight Service and get a weather briefing. We took off on Runway 27 and climbed straight out to the west. No chart was consulted. During the climb out, Hal made small talk, getting acquainted with his passengers. Flying the airplane was just something he did in the background.

There were clouds. We flew around them and kept climbing. At 6,500 feet, he leveled off. We were above patchy clouds and in smooth air. I would have flown beneath the clouds, because at this point in my experience, that's all I knew. But I had to admit that in spite of the heat on the ground it was comfortable in the cabin of the Cessna.

We motored on, still no reference to any charts, no checking for landmarks. Hal did not appear to be worried, but I was. I was the one who had to get us from Little Rock to Memphis, and then home again. I began to wonder if there was even a chart in the airplane.

Looking around, it dawned on me that I could see all four of the large reservoirs that dot the landscape in the northern part of the state. I could also see the Mississippi River ahead of us. All around were towns and highways that I recognized. Things looked a whole lot different from 6,500 feet than they did from the 1,500 feet I was used to flying. The air was also much more comfortable.

As we neared Clarksdale, Hal began to fiddle with the VOR and the next thing I knew there was a centered needle with an active "to" indicator.

"Little Rock," he said to me, pointing at the indicator.

We flew on with the needle centered. In a little while, he tuned the radio to the Little Rock approach control frequency (which it appeared he just happened to know) and gave them a call. With a slight throttle reduction, we started down.

Within a few minutes we were on the ramp at Central Flying Service in Little Rock. Hal gathered up his things and said goodbye and it was just Candee and me. She moved up to the front seat and looked at me expectantly.

Fortunately I found a chart in the seat pocket behind the pilot's seat and located the ground control and tower frequencies for Little Rock.

Finding Memphis wasn't hard. I just followed the Interstate out of Little Rock. It was one of my first IFR flights—"I Follow Roads." The ride was a little bumpier at 2,000 feet than it had been at 6,500—quite a bit bumpier. Still I elected to stay beneath the clouds, because that was what I had been taught to do.

Approaching Memphis, I consulted the chart for the tower frequency. Here I encountered my first challenge. The chart had different frequencies listed for Runways 9-27 and 3-21 than it did for Runways 17-35 left and right. I didn't know which runway I would be expected to use. Guessing, I tuned in the 9-27 frequency and called the tower.

"Memphis Tower, Cessna nine-eight-nine-eight-golf, fifteen miles west for landing."

"Cessna nine-eight-golf, Memphis Tower. If you're fifteen miles out, you need to be talking to approach control."

Oops. "Nine-eight-golf, roger."

I scanned the chart again for the approach control frequency. By the time I found it and got it tuned in I was over the Mississippi River.

"Memphis Approach, Cessna nine-eight-nine-eight-golf is over the river for landing at Memphis International."

"Nine-eight-golf, Memphis Approach, this is the wrong frequency. If you're on the west side you need to be on one-twenty-seven-point-eight."

"Nine-eight–golf, roger." This was getting embarrassing. I tuned in 127.8 and tried again. All the time we were getting closer to the airport.

"Memphis Approach, Cessna nine-eight-nine-eight-golf is approximately eight miles west of Memphis for landing." It was a guess, but the best I could do.

"Nine-eight-golf, this is Memphis Approach. If you're eight miles from the airport, you need to be talking to the tower on one-nineteen-point-three.

"Nine-eight-golf, roger." I switched again, wondering if I should just forget Memphis and turn south toward Oxford. I had flown in a

Terminal Radar Service Area (TRSA) before at San Antonio, but I didn't even know Memphis had one. I sure wasn't prepared for this flight.

I called the tower. "Memphis tower, Cessna nine-eight-nine-eight-golf, five miles west for landing." I imagined that the radar room had by now warned the tower to watch out for the idiot in the Cessna who didn't know what he was doing.

"Roger, nine-eight-golf. Winds zero-nine-zero at eight to ten knots. Report two mile final for runway niner."

"Nine-eight-golf."

Once we landed, finding my way around the Memphis airport on the ground presented its own set of problems, but I professed ignorance and told the ground controller that I wanted to go to Hi Air. He gave me directions.

The Memphis to Oxford segment of the flight was uneventful, because for that leg, I knew the way.

I'm not sure that I assimilated then the things that I learned that day. Looking back on it later, I realized my training had been very limited up to that point, with no real instruction on how cross country flying was done in real life. Take altitude, for example. By going above the scattered clouds, we accomplished several things. First, the air was calmer and more comfortable. Second, the extra altitude did wonders for navigation. Instead of looking at a small bit of the countryside and trying to match that with what was on the chart, we could see many miles and those miles were filled with major landmarks that made navigation much simpler. Radio navigation was easier, too. The extra altitude allowed us to receive the navigation radio signals over a much greater distance than would have been possible from a lower altitude.

Then there was the engine fuel-to-air mixture control. Since all of my airplane flying had been in a training environment and at relatively low altitude, I had never been taught how to lean the air-fuel mixture for a cruise flight. Hal Freeland had done this on the flight to Little

Rock, but I didn't know the procedure, so consequently didn't lean the mixture on my flight to Memphis, then Oxford.

By the way, Candee was real impressed with the flight. She slept most of the way.

Chapter 2 – The Vietnam Experience

I arrived in Vietnam the first week of October 1971, and my tour lasted until the third week of September 1972. The first couple of months in country were pretty busy. Then came a two to three month lull during which it literally seemed the war was over. Many of my flight school classmates who were flying "slicks" (Hueys used for transporting troops) or gunships were sent back stateside and offered early outs from their enlistment after spending just a few months in Vietnam. That wasn't happening with the medevac units, but that was all right with me. I wanted to fly as much as I could.

In April 1972, things began to heat up again and from then until the end of my tour, we pretty much had a war on our hands. Anyone who flew helicopters for a year during the Vietnam war has some stories to tell. Some of mine are fairly interesting.

Dustoffs at Night

My assignment in Vietnam was to the 57th Medical Detachment in the Vietnam Delta. It was an honor being associated with the 57th Med Detachment, which had a tradition of true heroes. Major Charles Kelly, the first Dustoff pilot to be killed in Vietnam, was the first commanding officer of the 57th and he really was a hero. Major Kelly was the first pilot to use the Dustoff call sign and was responsible for developing many of the tactics that were used by medevac crews throughout the Vietnam war. Another hero from the 57th was Medal of Honor winner Patrick Brady.

The first hint that my Vietnam flying would be done mostly at night was when my in country check ride was scheduled to start an hour before sundown. We did some autorotations, anti-torques, and hydraulics off landings before dark, then kept going after dark. Terry Greer, the IP who gave me the check ride, explained that the unit was in the process of transferring the bulk of its medical evacuation missions to the Vietnamese Air Force (VNAF), but the VNAF pilots wouldn't fly at night. So, the American Dustoff units were flying night medevacs while the VNAF flew most of the day missions. He assured me that the American crews flew all missions involving US troops, night or day.

Flying night after night opened up an entire new world to me. Often it was a lonely experience. Rarely were there any other helicopters in the air. The few fixed-wings that were flying, such as the Army's OV-1 Mohawk surveillance aircraft and the Air Force's Specter or Puff gunships, were usually on a different frequency. The Dustoff crew that had First-Up duty typically flew ten to twelve missions each night. Our only company was provided by Paddy Control, the Air Force radar controllers who provided radar coverage for most of the Delta.

One of the things you discover quickly when flying in third world countries at night is the absence of electric lights on the ground. Early in the evening there are numerous cooking fires around the countryside, but later there is nothing once you get away from the major towns. It's mighty dark down there.

The techniques we used for locating the troops on the ground provided an interesting set of challenges. As a new pilot in country, I learned these techniques from the more experienced aircraft commanders. Staying alive and successfully accomplishing the missions depended upon the Dustoff crews developing excellent crew coordination skills. Everyone was depended upon to do their part.

Our mission coordinates were plotted on a 1:50,000 tactical map. Paddy Control could normally provide vectors to the general vicinity of the pickup site. From there, a safe landing could only be assured if two-way radio communication with the troops on the ground was established. Some of these troops were very experienced at helping helicopter pilots find a remote pickup site and some weren't. Many of the Vietnamese units had at least one American advisor who could handle communication with the Dustoff crew. For other missions, a Vietnamese interpreter flew with us and handled communication with the elements on the ground.

A variety of methods were used for identifying the landing area. First contact was normally made by sound. We flew without external lighting so it wasn't easy to see us. There being no other aerial traffic around, there was no reason to provide a visual target for an enemy gunner. But, to the guys on the ground, the sound of an approaching Huey was unmistakable. Upon first radio contact, we would confirm our proximity to the landing site by asking the ground troops if they heard us, and if so which direction were we from them. A typical response would be along the lines, "We hear you, Dustoff. It sounds like you're north of us about two klicks." In that case we would turn south and ask the ground contact to tell them if it sounded like we were getting closer.

Once in visual range, we needed some type of light to identify the landing area. Among the possibilities were a hand-held strobe light, a flashlight, a flare, a small fire, or four small fires built to form a "T" on the ground. Most of us preferred the latter, because it gave us not only a marker for the landing site, but something to line up on for our approach. The base of the "T" was oriented along the recommended approach path.

The least preferred method to mark a landing spot was a flare. A flare had the dual disadvantages of highlighting the position of the ground troops and adversely affecting the night vision of the Dustoff pilots, but sometimes that was all they had.

More often than not, a handheld strobe light or a flashlight aimed toward the helicopter were what we had to work with. The radio operator on the ground would describe the landing area to us, including any obstacles and an estimate of the wind direction and velocity. They would also recommend a flight path back out, based on obstacles and location of any known enemy. Rarely did it make sense to go back out the same way we came in, but sometimes that was the only choice.

Using whatever we had to work with visually and the landing area briefing from the contact on the ground, we would set up a rectangular landing pattern. This would help us stay oriented with the landing spot. We made our approach to a point just shy of the landing marker. The pilot not flying called altitude and airspeed every few seconds as the aircraft descended. He also kept his hands near the controls. Some Aircraft Commanders even preferred that both sets of hands be on the controls during short final. The landing light was delayed as long as possible for obvious reasons. Without the light, enemy gunners had only sound and a possible shadowy outline to aid in target acquisition. Once the landing light came on, the Huey was a well-lit target.

The approach was made all the way to the ground because of dust and debris, but we would hold the chopper light on the skids. Who knew what kind of surface you were landing on? It could be muddy, wet, lumpy, or littered with stumps or stubs that could easily puncture a hole in the bottom of the aircraft, and more importantly, the fuel cells.

Throughout the approach, the medic and crew chief hung out the sides of the aircraft watching for obstacles or enemy activity. Constant chatter on the intercom was a necessity to keep the pilots informed, since we couldn't possibly see all that was around and beneath us.

At touchdown, the landing light was switched off and the crew in back supervised the loading of patients. They were not to unplug from the intercom in the event a rapid departure was needed. The helicopters were equipped with extra-long microphone cords that allowed the medic and crew chief some freedom of movement outside the aircraft.

Seldom did more than fifteen or twenty seconds pass before the crew reported being ready to depart. When the patients were on board, the medic would say, "Ready right" followed by the crew chief with "Ready left."

The pilot flying would announce, "Coming up."

And from the back, "Clear up left."

"Clear up right."

The landing light was used on departure only if necessary to insure obstacle clearance. Otherwise it was best for the pilots to regain their night vision as soon as possible.

It was not unusual for the pilot flying to be coached by the other pilot. "Watch your torque, you're at 38, 40, okay nose her over."

The crew chatter would continue, "We're still clear left."

"Clear right."

And off we'd go, either to a small local hospital, or perhaps another pickup. If there were Americans on board, we would head for the nearest US Army hospital. En route, the medic and crew chief would be busy treating the patients. This scenario would be repeated multiple times each night, night after night.

Two of Everything

Date: 11/4/71
A/C Type: UH-1H
Registration #: 69-15015
Route of Flight: Navy Binh Thuy – Vung Niem – Mo Cay – Dong Tam – Ben Tre – My An – My Tho

1Lt. Steve Hamman was the First-Up Aircraft Commander. Though he had a seasoned crew in the back of the aircraft, he was saddled with me as "peter pilot." I had barely a month in country and had been on the duty roster only a couple of weeks. When we walked into Operations after an early dinner and saw the mission board, we knew it was going to be a long night. The three VNAF crews had been humping all day and still the mission board was full. Charlie had been busy in the Delta.

Steve copied down the mission coordinates while I went out to preflight the Huey. When Steve came out a few minutes later, I was just climbing down from the top of the helicopter, having completed a visual inspection of the main rotor system. The medic was busy securing the litters and checking his medical supplies, while the crew chief made his own inspection of the aircraft and logs. The Huey had just been through a periodic inspection that afternoon.

"We'll get some action tonight," Steve said as he hung his M-16 on the back of his armor-plated seat and zipped up his survival vest. "All of the missions were called in as *secure*. Odds are, at least half of them will be *hot*."

Even as a newby, I knew what he meant. It was just one of the nuances of war that the Dustoff crews had learned to live with. If a mission was called in "hot" that meant the troops were engaged with the enemy, or as we called it, "in contact." Army regulations required that Dustoff helicopters have gun cover whenever the troops were in contact. Gun cover was supplied by Cobras from Vinh Long or Can Tho (*Can Toe*), or the Navy Seawolves in their Mike Model Hueys.

A ground commander in contact knew that it took time to coordinate the gun cover between units. The gunships may be committed to another mission, or, as in this case, when the mission was at night, gun crews weren't normally available without at least an hour's notice. If you really needed gun cover under these circumstances, by the time you got it, it would be too late. If the ground commander was concerned about his troops and needed to get them out in a hurry, calling the mission in as "secure" was the only way to insure a rapid response from the medevac unit.

If the AO really was secure and the patient wounds weren't critical, it was a different story. The ground commander would figure, "We've got time. We might as well let Dustoff get some gun cover, just in case." So, the hot missions were often called in as secure and the secure missions were often called in as hot, and that's just the way it was.

We loaded the aircraft with our personal weapons before strapping into our seats. These weapons were a varied assortment, according to personal preference. All the crew weapons we carried were "off books." Our assigned weapons were stored in the arms room, where they stayed because there was too much paperwork involved if an assigned weapon was lost or stolen.

I had an M-2 carbine, an M-16 taken off a dead ARVN, and a .38 Special Smith & Wesson that had been given to me by one of the ACs that had DEROSED. Steve Hamman flew with an M-79 grenade launcher and a .45 caliber "grease gun" (a machine gun that might have been used by a gangster), in addition to his "off books" M-16. The medic and crew chief each carried an AK-47 and an M-79. (Remember, medevac helicopters weren't "armed.")

Our first mission of the night was near Vung Nem. It was an area that had been quiet for a while, but during the last few days the VC had been ambushing local ARVN patrols and raiding the villages.

As we approached the coordinates we had been given, we contacted the ground unit on the radio. The American advisor with the ARVN patrol advised that they had been ambushed a few minutes earlier. They were currently pursuing their attackers, who had fled to the south. He advised us to approach from the north to avoid enemy contact. He indicated he was receiving and returning sporadic small arms from the south.

The ARVNs were in a tree line bordering a small field. One of the ARVNs stepped out of the trees holding a battery-operated strobe light. I was at the controls; Steve was working the radio. We approached from the north, and landed in the field, just short of the strobe light. Soldiers ran out of the trees carrying five wounded ARVN soldiers and helped load them on the helicopter.

Less than 30 seconds after we touched down, the crew chief called, "Let's go." The medic was already bending over one of the patients, trying to seal a sucking chest wound.

I picked up the helicopter, backed it away from the trees, and made a 180 degree turn about the mast. Dipping the nose, I raced low-level across the field, picking up speed. Just before reaching the trees on the other side of the field, I pulled back on the cyclic and we zoomed into the air. A few shots were fired at us from beyond the trees, but nothing really close.

We dropped the patients off at the Vietnamese clinic in My Tho, and departed with enough fuel for another mission.

Paddy Control vectored us to a site just north of Vinh Long. The RTO with the ground unit advised they were in the midst of a firefight, and had three critically wounded requiring immediate evacuation. He indicated he was holding a strobe light to mark the landing area. We looked for the strobe and when we found it, there were two of them about 200 yards apart. We advised the RTO that we had two strobes and he told us he would shine a flashlight at us. Within seconds we had two flashlights beaming up at the aircraft. When we advised the RTO that the bad guys were monitoring his frequency, he said he would build a small fire to help us identify the landing site, which was on the north side of a small canal that ran east and west. Immediately we saw two small fires along the canal. No tracers were visible at this time, indicating a lull in the fighting.

Steve was flying now. I was working the radios. "I've got two small fires, Red Dog," I radioed. "Which are you?"

"We should be the one nearest you, Dustoff. It sounds like you're right on top of us."

We weren't on top of either fire, but it probably seemed that way to him. A Huey can be awfully loud, especially at night when you can't see it. By this time Steve had descended to about 200 feet and was flying slowly along the canal. We were totally dark, with all outside lights off. Steve indicated he didn't feel too good about the situation. I didn't either.

"Ask him if he's the east fire or the west fire," Steve told me. I relayed the question.

"I don't see another fire, Dustoff, but if you've got two, we should be the one to the east," Red Dog replied.

We were almost on top of that east fire and Steve lowered the collective. The whole crew was uneasy, and we expressed as much to each other. Something just didn't feel right. Steve told me to flip on the landing light. As soon as I did, all hell broke loose. Our windshield filled with green tracers, red tracers, and orange tracers. It looked like a fireworks display. Miraculously they all converged at a point twenty or thirty yards in front of us. My immediate thought was that it was like we had a giant invisible shield in front of us.

I turned off the landing light and Steve altered his approach to land at the other fire. He keyed the mike switch on his cyclic and calmly advised, "No, Red Dog, you're the west fire."

We made the pickup and departed to the north without taking any more fire. We left the three patients from that pickup at Vinh Long, landing at the river helipad, where an ambulance met us. We then went to Dong Tam to refuel.

The refueling was done "hot" since Dong Tam, once a sprawling American base, was now left with only a few ARVNs to defend it.

We made several more pickups as the night progressed, including some civilians from a little village. The VC had come through and shot up the place, including women and children. We took these casualties to the My Tho (*Me Toe*) hospital. When we stumbled into breakfast around 7:00 a.m., we were one tired crew. We had flown eleven sorties during nine and a half hours of flight. The total patient count for the night was thirty-seven. All survived.

Attack on Tieu Can Base Camp

Date: 11/17/71
A/C Type: UH-1H
Registration #: 68-16200
Route of Flight: Navy Binh Thuy – Muc Hoa – Tra Vinh – Tieu Can – Can Tho

The first attack on the base camp outside of Tieu Can (*Too-E-Kän*) came just after midnight. It started with mortars—extremely accurate mortars. The first one landed inside the compound and left two ARVN soldiers dead. A few minutes later, the sniper fire started. The VC sniper obviously had good night vision, perhaps even some type of night vision scope. His firing into the compound was extremely accurate.

Several of the ARVNs slipped out of the camp and ran away prior to the first VC attempt to bridge the perimeter. This attempt was an assault on three sides simultaneously. The American advisor used flares to light up the area so that he and the remaining ARVNs could pick the VC off as they made their way through the concertina wire. There weren't enough ARVNs to cover all three sides at once and two of the VC made it into the compound with hand grenades. By the time they were killed, eight more ARVNs were dead. Several more left the compound as soon as the flares burned out.

The American advisor called his commanding officer in Vinh Long and requested an evacuation. He told his commander that the gooks were coming through the wire and the ARVNs were deserting him. The C.O. promised to get him some help ASAP. His first action was to contact his Vietnamese counterpart to see if there was an ARVN patrol in the area that could be diverted to the outpost. The ARVN commander checked his situation map and noted that the nearest patrol had just left

Cau Ke (*Cow Key*), over twenty kilometers away. He started them toward Tieu Can, but it would take them at least four hours to get there.

Meanwhile, the American advisor at the ARVN outpost advised that he had only seven ARVNs left with him. The rest were either dead or had deserted. Ammunition was also in short supply. The C.O. promised to find a helicopter.

He called the 188th Assault Helicopter Company in Can Tho. He couldn't reach the commanding officer, but was told by the night clerk there were no crews on duty. "This is an emergency!" the C.O. pleaded. There was nothing the specialist could do without finding his C.O. or the Operations Officer.

Desperate to save his man, the C.O.'s next call was to Dustoff Operations. He called the mission in clean, telling the RTO just exactly what was going down. Without hesitation, the RTO called the First-Up crew on the radio. Gary Chester was the Aircraft Commander. I was the peter pilot. We had just dropped off a load of patients at Phu Vinh. That put us less than ten minutes flying time from Tieu Can.

Gary flew in that direction, while I copied down the frequency and call sign of the American Advisor at the Tieu Can outpost. I switched the FM radio to the new frequency and gave the guy a call. "Roadrunner two-three, this is Dustoff seven-eight."

"Dustoff seven-eight, this is Roadrunner two-three," came the immediate reply. There was just a hint of apprehension in his voice.

"I hear you have company down there." I was still a newby, but already I had learned the technique of understatement. If you listened to our voices on the radio, we never encountered a situation that was too difficult to handle, or caused us to be afraid. (Roger that, Ops. We've lost a rotor blade and the engine's on fire. We'll be back with you in a minute—sort of the Chuck Yeager approach to life in the danger zone.) Maybe we felt it was our responsibility as pilots to present a calm, macho image in the midst of danger. This attitude tended to convince us we could do anything, which is how we managed to save lives.

Roadrunner picked up on it right away. "Roger, Dustoff. We've got a few uninvited guests down here," he replied.

"Are you ready to leave the party?" I asked.

"I could be convinced, Dustoff, if you could find it in your hearts to stop by and give me a ride." The guy was probably scared out of his gourd, but since we hadn't let on to him how scared we were, he couldn't let on to us how scared he was.

Soon we were close enough to see the action. Tracers were flying in both directions, but more were going in than were coming out. The incoming tracers were mostly green from AK-47s. The outgoing tracers were red M-16 tracers. If there was an M-60 in the compound, it was either out of ammo, or there was no one to fire it.

"Can you get to a point where we can pick you up?" I asked Roadrunner.

"On the blue side, Dustoff," he responded. That indicated he would be coming out of the compound on the south side, near the canal.

"Roger, Roadrunner. Keep your head down, and give us a light when you can," I told him.

"Wilco," was his only reply.

Gary maneuvered the helicopter so that we could make an approach along the canal to an area just outside the camp. He started a descent before we saw the strobe. Our external lights were still off.

There were four of us in the helicopter and we were all on the edge of our seats. Our eyes were moving constantly, trying to take everything in. My hands were near the controls. We expected the VC to have been monitoring our radio conversations. They would be moving to cut off those fleeing the camp and hoping to bag a helicopter in the process.

I wanted to know how many we were picking up, but knew better than to ask. If Roadrunner told us, he would further weaken his position by giving the enemy his strength. We asked the crew if they were ready. AK-47s in hand, they both assured us they were. What we wouldn't have given for a Seawolf or two about now.

When we touched down, the American Advisor and four remaining ARVNs were crouched in some bushes waiting for us. The scream of that turbine engine and the throbbing of those big rotor blades probably felt to them like they were being run over by a freight train. They told us later it was the sweetest sound they had ever heard.

When the sound was right on them, the American switched on his strobe light. I didn't see it and neither did Gary. Pete Petersen, the medic, saw it first, almost directly below us. He called out the position and Gary made a hard left turn while I switched on the landing light. Immediately tracers came our way, but they were shooting behind us, misjudging our forward motion because of the darkness.

We were kicking up a lot of dust and a roll of concertina wire was blowing around in the rotor wash, so Gary took the helicopter to the ground. In the back, Pete the medic, and Eddie our crew chief, started laying down covering fire with their AK-47s, spraying the area where most of the incoming fire was originating. The American advisor and the four ARVNs dove for the open door on my side of the helicopter. Eddie signaled for us to go. Three rounds shattered the plexiglass in the small, outward swinging door that was just behind Gary's seat. I was already on the controls with him and it became my aircraft as he reacted to the impact of the rounds. I was pulling pitch and pedal-turning away from the firing at the same time. The landing light was still on, so I flipped it off and dumped the Huey's nose. Gary was back on the controls, but continued to let me fly. As we climbed for altitude, we escaped the small arms fire and things settled down a little. It's a good thing we were both still on the controls, however. The American sergeant we had just picked up was leaning over the console, trying to hug us both.

The next morning on an ash and trash (slang for administrative flight) run to Vinh Long, we flew over what had been the Tieu Can base camp the night before. There was nothing left but a blackened spot on the ground.

Bullfrogs and Runways

Date: 1/2/72
A/C Type: UH-1H
Registration #: 68-15015
Route of Flight: Navy Binh Thuy – Vung Liem – Cai Le – Can Tho

Rick Slade, the 57th's Maintenance Officer and I were coming back from a mission late on the night of January 2, 1972, only to discover Navy Binh Thuy (*Ben Too-e*) totally engulfed in a heavy rain shower. Flying down the Bassac River, we were in the clear, but couldn't even see the lights of our home base. Running low on fuel, we opted to head further down the river to Can Tho Army Airfield.

The tower operator at Can Tho reported heavy rain, but as we approached, we could see the runway lights. I found that interesting because the runway lights at Can Tho were not electric, but oil-filled smudge pots that were lit every night. I don't know why the rain didn't put them out, but it didn't.

Approaching the runway, we flew into such a heavy downpour that all forward visibility instantly went to zero. Rick was flying the aircraft. When his visibility went to zilch, he had me get on the controls with him. Neither one of us was instrument qualified.

Rick had me turn on the landing light. The landing light on a Huey is retracted into the belly when you're not using it. When you turn it on, you have the ability to adjust it to match whatever approach angle is appropriate for the situation. This time, as the light started moving forward and while it was still pointing almost straight down, I told Rick I could see the runway below us.

"You take it then and hover us straight down."

He never actually got off the controls, so it was a combined effort. Needless to say, we were aided very much by the flight crew, who in spite of the rain opened the sliding doors on either side of the aircraft and helped to talk us down.

It was basically a "hover down" from two hundred feet above the runway. We didn't have much depth perception. What little we had was aided by the smudge pots on either side of us and the fact that there was a white line painted down the center of the runway.

Ever so slowly, we eased down, down, down, literally feeling for the ground. When the skids did touch down, we both pushed the collective pitch down forcefully, as if to stick us to the deck. We dared not try to hover to a parking place, so we just sat there. Suddenly, I laughed and pointed. There in front of the helicopter, hopping across the runway was a big old bull frog.

VC Cousin

Date: 1/26/72
A/C Type: UH-1H
Registration #: VNAF-585
Route of Flight: Navy Binh Thuy – Dong Tam

Ron Ihler, the new Operations Officer, summoned me into his office. "Well, Freeman," he said. "You're a new aircraft commander. Guess what you get to do tonight?"

I was Third-Up on the duty roster. Normally, Third-Up didn't do anything but ash & trash, and they didn't do those at night. "I don't know. What do I get to do tonight?"

"Fly with the VNAF."

I knew some of the older ACs had flown with the VNAF Dustoff crews before I got to Binh Thuy, but they hadn't been doing it lately.

"What do you mean, 'fly with the VNAF'?" I thought he was just toying with me.

"You get to fly right seat with one of their ACs," Ron explained. "We're supposed to be pulling out of here in a few weeks, and Group wants us to make sure those guys can handle their own missions at night."

"I didn't think it was a question of could they, but a question of would they," I replied. I didn't know the politics behind the decisions, but I did know that we flew medevac missions night after night, while the VNAF Dustoff ships sat on the ramp. Ninety-nine point ninety-nine per cent of our patients were Vietnamese.

Ron seemed as dubious as I did, but he had a job to do. "I don't know what's going on any more than you do," he explained, "but I do know I've got to put a pilot with the First-Up VNAF crew tonight and you're it. The rest of us have already had our turn in the barrel."

"Okay," I said. "Where and when?"

"Meet Lieutenant Phouc (*Fook*) on the west side of the runway at nineteen hundred hours."

I knew Phouc. I'd heard he had about eight thousand hours and was an excellent pilot, though he could care less about radio and traffic pattern procedures. We joked about it. "We come, we go," seemed to be the extent of VNAF communications with the tower.

Nineteen hundred hours military time was seven p.m. Warrant Officer time. When my new Seiko indicated it was time, I crossed the runway and waited for the VNAF ship to arrive. He was about an hour late. When he landed, his right seat was occupied by a grinning crew chief who jumped out and held the door open for me to get in. Phouc did not shut down. That meant I wouldn't have the opportunity to do a preflight inspection. It was a VNAF-owned Huey. I wondered about their maintenance. I wondered if they even did a preflight.

We took off and headed north. Except for the "we go now" call to Navy Binh Thuy tower, all the radio conversations were in Vietnamese. Needless to say, I was more than a little uncomfortable when a half an hour later we began circling over a firefight a few klicks east of Dong Tam. Judging from the tracers flying back and forth on the ground, and the abundance of high-pitched, and obviously excited radio transmissions, none of which I understood, there was quite a battle going on down there. I didn't have a clue what Phouc was going to do, and he was so busy on the radio, I couldn't break in to ask him. Some "advisor" I was turning out to be.

I noticed Phouc checking the gauges and stretching out a downwind leg like he was going to land. I didn't like what I saw. Finally, Phouc took a break from the radio and looked over at me. In the dim, red glow of the cockpit lighting, I could swear he was grinning from ear to ear. I was determined not to let him know how uneasy I felt.

"What's going on?" I asked him.

"It's my cousin down there," he replied. "He's begging me to come down."

That I could understand. My own cousin had been a door gunner in Vietnam a couple of years earlier. If he had been in trouble and if I had been around then, I would have gone in to get him, firefight or not. "Let's go get him," I said, trying not to let him hear any signs of fear in my voice. If we were going into the midst of a firefight, at least we would be going after someone that mattered to someone in our crew.

"Oh, no, no, no, no," Phouc said, laughing. "That's not the deal. My cousin is VC. He wants us to come down there so he can B-40 us." A B-40 was a rocket. It did bad things to helicopters.

I had been around Vietnamese people enough to know that I didn't understand them. They would pass up a chance to save a wounded soldier if it was the least bit risky, but wouldn't hesitate to pick up a KIA, regardless of the risks. I didn't know if Phouc was going to take the challenge or not. He seemed to be having too good a time to suit me.

"So, what are we going to do?" I asked him.

"Don't worry," Phouc replied. "I told him to wait. We will be down in a couple of minutes. I didn't tell him we have Cobras coming. We're going to blow him away before he blows us away."

Obviously, Phouc was enjoying himself. I began to relax. It sounded like he had a pretty good plan. But where did he get Cobras at night? I had never been able to do that. I wondered if VNAF had Cobras. If they did, I had never heard anything about them.

The situation reminded me of kids playing war. Only these guys had real weapons, and they were playing for keeps. "Can't shoot through bushes!" "I got you!" "No, you didn't!" These rules wouldn't work here. I wondered if there were wounded down there, or if Phouc was just playing games with his cousin. Not that I could do anything about it. Phouc was commander of the aircraft, and I was his peter pilot, even if I was supposed to be teaching him the techniques we used for locating and landing at suitable PZs at night. He and I both knew that he had far more experience at this than I ever hoped to have.

We pulled a little away from the action and waited for the Cobras. Phouc had me turn on the rotating beacon (first thing I had gotten to do all night) so the Cobras could spot us when they arrived. We didn't have to wait long before a couple of Outlaws showed up and hailed us on our frequency. They were Americans. I recognized the call signs. They had covered me before on daylight missions, but they rarely flew at night. They were pretty vocal about not liking it, too.

"Dustoff, this is Outlaw 22. ID us a target and let's get this show on the road."

Phouc turned to his Texas-Alabama vocabulary for this part of his mission. He identified targets for the Cobras by the muzzle flashes of the weapons being fired on the ground. The lead Cobra pilot allowed as how it would be pretty hard to fire that close to the friendlies, since he and his wingman weren't allowed to go below a thousand feet at night.

A thousand feet?! I didn't have any sympathy for them at all. I flew eight to ten hours a night and seldom got *above* a thousand feet.

With Phouc guiding them, the Cobras went in and let go one salvo each, then Phouc set us up to make the pickup. The Cobras hung up high, assuring us that if there was any sign of firing at us, they would neutralize it. I figured that if Phouc's cousin was still down there with his B-40, it would be all over before the Cobras could react. I hunkered down in my seat and adjusted my chicken plate, as if it would offer any protection against a B-40 rocket.

Phouc lowered the collective and started bleeding off airspeed. I kept my eyes peeled for any sign of tracers headed our way. My hands were near the controls, ready to take over at a moment's notice. There was still some small arms fire on the ground, but it was becoming sporadic. A small fire marked our landing spot.

I wanted the controls, wanted to be in control of my own destiny, but it was Phouc's aircraft and he didn't offer them to me. There was no doubt that he was a better, more experienced pilot than I was, but dying while flying copilot to a Vietnamese national was not among the headlines I had imagined describing my death.

We made it to the ground without getting blown out of the sky. As we touched down, a light flared off to our right and a loud "whoosh" seemed to take all the oxygen out of the air. My side of the aircraft looked like the fourth of July. It was as bright as day, destroying any semblance of night vision. I saw one Cobra, then another swoop past and begin a steep climb. They had obviously gone below a thousand feet. It looked like they were coming back around for another pass.

"What happened?" I wondered aloud.

"I think they got my cousin," Phouc said, matter-of-factly.

I heard the Outlaw pilots congratulating each other on the radio for what they had done. "I don't think you'll have any more trouble with B-40s, Dustoff," Outlaw 22 told us. "We'll wait until you're out of the hole before we head for home."

"Roger, roger," Phouc answered them. "We're coming out now. Thanks." I realized there had been activity in the back of the aircraft and looked over my shoulder to see that the medic and crew chief had managed to fill the cabin with wounded ARVNs while all the action was going on.

"Don't mention it," Outlaw 22 said as they flew off toward Vinh Long.

"You've got it," Phouc said to me, sounding every bit like a Texan.

I put my hands and feet on the controls, surprised I was finally getting to fly. My eyes were still trying to adjust to the darkness again. "I've got it," I said, and pulled pitch.

There's an interesting side note to this story. Months later, after the 57th had moved to Long Binh, I was flying in the Delta and landed one day at the Can Tho soccer field to drop off a couple of patients. A Vietnamese officer came out to the helicopter and asked us to shut down. He indicated there was a patient in the hospital who wanted to talk to me. Since it was a Vietnamese hospital, I was a little surprised, but curious. We shut down and followed him inside.

There, lying in a bed in the middle of a ward that was packed with patients, was Lieutenant Phouc. His eyes brightened up when he saw me. "Freeman! Am I glad to see you."

"What's happening, Phouc?"

"It's these crazy Vietnamese doctors. They want to take off my leg," he told me.

"Really?" The thought of that happening made me queasy. "Why?" I asked him.

"They're butchers," he replied. "That's all they know how to do." This was from a Vietnamese national, talking about doctors from his own country. I wasn't sure what Phouc wanted me to do about it.

"Freeman," he pleaded. "Take me with you to Saigon, to an American hospital, where they can save my leg."

"I don't know about Saigon," I told him, "but maybe I can get you in the 24th Evac at Long Binh. I'll have to make some calls."

"Take me with you, now," Phouc begged. "If you leave me here, it might be too late."

I didn't know where to start with the red tape, but figured if we took Phouc to the 24th Evac, they would take care of him. I figured wrong. Since we weren't scheduled to return to Long Binh for a few days, I couldn't take Phouc right then, anyway. We went over to Can Tho Airfield and I got on the land line to Long Binh. Three hours later I had an answer. The 24th Evac couldn't take him, but an American doctor there had expressed an interest in coming to Can Tho to see what he could do.

Arrangements were made, and as far as I know, the American doctor did make it to Can Tho to perform surgery on Lieutenant Phouc's leg. I hope he was able to save it.

In May 1997, I learned that a former VNAF Dustoff pilot, Trang Van Phouc, was in the United States and living in the San Jose, California, area. At one time he had an e-mail account and was in contact with some of the American helicopter pilots that communicate through the Internet. I tried calling information but could find no listing. I tried contacting the VNAF web site and didn't locate anyone there that knew him. If anyone reading this knows Trang Van Phouc and can put me in touch with him or him in touch with me, it would be interesting to know if this is the same man and whether or not his leg was saved.

Keeping a Mohawk Company

Date: 2/5/72
A/C Type: UH-1H
Registration #: 69-15223
Route of Flight: Navy Binh Thuy – Rach Gia

While in the air on the night of February 5th, 1972, we got a call from Paddy Control telling us he had an emergency working near Rach Gia (*Rock Jaw*). He asked if we could help out. Temporarily between missions, we told him we would be glad to.

An Army OV-1 Mohawk had been flying a reconnaissance mission near the western coast when one of his turboprop engines had failed. The crippled aircraft was heading toward Air Force Binh Thuy and was losing altitude. We turned to a heading to intercept him via radar vectors and contacted him on the frequency given to us by Paddy.

"Eavesdrop, three-nine, this is Dustoff seven-four," I radioed.

"Dustoff seven-four, it's good to hear your voice," the Mohawk pilot replied. It was a dark, lonely night over the U Minh Forest. We were the only other aircraft around.

"Roger, Eavesdrop. We're headed your way. Thought we might tag along with you to Binh Thuy, if you have no objections."

"Objections, hell. It was getting kind of lonely out here, Dustoff. I'm down to about 6,000 feet and maintaining altitude, but they don't call these planes 'widow-makers' for nothing."

I knew how he must have felt. There had been a number of night accidents in Vietnam and most of the night flying, except for some special missions, had been eliminated. Our mission required us to be in the air nearly every night. Most of the time we were the only ones flying in all of the Delta. It was a lonely feeling and only the watchful eye of Paddy Control's radar kept it from being totally frightening. The Eavesdrop Mohawks, flying their surveillance missions were probably around, but we never heard them because they flew at high altitude and talked with the controllers on a different frequency.

Within a few minutes, we got a visual sighting of the OV-1's rotating beacon. We stayed in contact with him until he was on the ground at Air Force Binh Thuy. He would have done the same for us.

This is another story with a sidebar. On October 15, 1999, I received the following email in response to the above story being posted on my web site:

To Dustoff Seven-Four

Just got back on the 131st all unit listing and saw your note about "Keeping A Mohawk Company".

Unless I miss my bet I am that Mohawk Pilot. I don't remember the specific day, but it was in Feb. '72 and I was flying a mission that took me from Long Than (73SAC) Base In a triangle route out over Laos and down over the U-Minh and then back to Long Than. I was at mission altitude of 10,000 feet and turning on my leg to the U-Minh when my right engine went out. I cleaned up the aircraft and according to how we were taught in flight school should have been able to maintain altitude with one engine.

For some reason not known, this aircraft would not maintain the altitude. You said I was at 6,000 feet, but I had in fact gone down to 3000 feet. I told my TO that if we went any lower than 3000 feet we would punch out. I called Binh Thuy and asked if they could send a Dustoff my way. Was I ever glad to hear your voice.

I made a safe landing, returning a couple days later to Long Than, and then later went on to fly for the 131st out of Marble Mountain, with missions off the coast of North Viet Nam.

I don't remember saying thanks before, so I'm saying thanks now.

Paul Weisenberger

Night Hoist Mission From a River Boat

Date: 5/23/72
A/C Type: UH-1H
Registration #: 69-15564
Route of Flight: Long Binh – Go Da Ha

I got to go home to Mississippi for a couple of weeks at the mid-point of my Vietnam tour. The Army paid for seven days of R & R in Hawaii. I added a week's leave to that and paid my own way from Hawaii to Mississippi and back. On my second day back from R & R, I drew First-Up night. Home to Mississippi, then back at the war, just like going to the office. At least I wasn't on the daytime shift. I wasn't quite ready for Lai Khe (*Lie Kay*) standby and picking up wounded ARVNs in and around An Loc.

Because I'd been gone for two weeks, Captain Jackson, the 159th's CO and an IP gave me a Standardization check ride during the afternoon. Fortunately, he had the foresight to throw in a little refresher training on hoist missions.

The crew gathered at the aircraft half an hour before our shift was to begin so we could do the preflight and runup. Then we headed for the crew lounge to shoot some pool. Night times were pretty quiet since we'd moved to Long Binh—nothing like the all night flying we'd been doing in the Delta prior to the move.

I was comfortable with my crew. The peter pilot, David Smith, was still a newby, but had a good head on his shoulders. Tom Pierce was the crew chief, and Mike Toomey the medic. Except for Smith, it was a crew I had flown with many times. We played a few games of eight ball, and Toomey decided he wanted to go to the ready room and get some sleep.

It was almost midnight and the rest of us were about ready to turn in when the phone rang. All three of us just looked at it at first. Then I picked it up. The RTO's voice sounded a little excited. "Dustoff, sir," he said.

Tom didn't have to wait for me to tell him to go get Toomey. He had already headed toward the crew lounge as soon as I picked up the phone.

Smitty and I ran to the aircraft. I untied the blades as he climbed in and strapped himself in. He was yelling "clear" and pulling the starter trigger as I climbed in the left seat.

The crew was with us before the rotor was in the green. Smith lifted off as I called Ops on Fox Mike. "Dustoff Ops, this is seven-four, we're off. What have you got for us?"

"It's a river boat, seven-four. Let me know when you're ready to copy the coordinates."

I pulled my plastic-covered map out of the leg pocket of my flight suit and reached for the grease pencil that was in my left shoulder pocket. "Go Ahead," I transmitted, using the foot mike switch on my side of the aircraft.

"Coordinates are x-ray-tango-one-zero-four-four-zero-zero. It's just south of Go Da Ha on the Vam Co Dong River."

"Copy," I replied. "Who's the contact?" I pointed to the northwest to give David the direction to head before turning my attention back to getting the details of the mission. The lights of Saigon were bright off to our left. Visibility appeared good, but above us the sky was overcast.

"Contact River Rat Three on thirty-eight point twenty-five."

"Roger."

I moved the ICS selector up one setting and contacted the Air Force radar controller on UHF. "Paris Control, this is Dustoff 74, off Long Binh, headed for x-ray-tango-one-zero-four-four-zero-zero, squawking 1200." I glanced quickly at the transponder and noticed we weren't squawking anything. I moved the switch from "standby" to "on."

It took a minute before Paris came back. Meanwhile, I was looking for the FM frequency for the artillery advisory for Phu Loi (*Foo Loy*).

Before I found it, Paris advised me, "Dustoff seven-four, you're radar contact three clicks northwest of Long Binh. Looks like three-five-zero for three-five miles will get you to your coordinates."

"Dustoff seven-four, roger," I glanced at the RMI to make sure Smith was picking up the heading, then dialed in the FM frequency we needed to get an artillery clearance.

I quickly ascertained that there was no artillery firing along our course and I switched back to the Dustoff Ops frequency to see what else I could find out about the mission. The RTO was just full of good news. "They were disabled in an ambush, seven-four. When they called the mission in, they were still taking fire from one of the river banks. They've got two critically wounded."

"Any chance for some guns?"

"Are you kidding?"

A few months earlier and we'd have had help from the Seawolves. There were Cobras at Bien Hoa (*Ben Wah*)—Blue Max. They were taking a beating, compliments of An Loc, Loc Ninh, and Tay Ninh, and weren't putting up crews at night for the time being. There might be some Sabre guns available. I knew the RTO would be trying all his options. Meanwhile, we flew towards Go Da Ha, lights out at 2,000 feet, with no idea of what to expect when we got there.

We passed north of Cu Chi and Trang Bang and I tried to pick up the river in the dim light. A little moonlight would have been nice, but it wasn't to be. Paris told us we were getting close and needed to turn a little more to the north. David turned ten degrees to the right and I tuned in the tactical frequency we'd been given. Almost twenty-five minutes had passed since we'd left Long Binh.

"River Rat three, this is Dustoff seven-four."

"Dustoff seven-four, this is River Rat three." Was that a fifty cal I heard firing in the background?

"We're about five minutes out. What's your tactical situation?"

"Not good, Dustoff. We're adrift and taking fire from the river bank. The engines got hit with an RPG. We're taking water, and I've got four wounded, two critical, over."

Not good, was right. They were taking fire, they were in the river, and we had no gun cover. I wasn't sure how this was going to pan out.

"Can you mark your position for us?" I asked.

"Affirmative, Dustoff, but we're still in the middle of the river. We're trying to get downstream where we can get ashore, but we have no engine power. Plus, we've been a little busy."

I heard what came out of my mouth next, but didn't believe I had said it. "That's okay, River Rat three. We can use the rescue hoist to get your wounded if we can locate your position." Maybe it was because we had done some hoist training earlier in the day. It certainly wasn't because I had a lot of experience with hoist missions. In fact, I'd never done a real one, except to sit and watch as a Peter Pilot. I'd never heard of anyone doing one at night. I heard and sensed activity behind me as Pierce and Toomey began to rig the hoist.

I still didn't see the river, but suddenly I saw tracers ahead—red ones, big red ones. There were green tracers coming from the opposite direction.

"We've got you in sight," I told River Rat three. "Can you give us some lights for positioning?"

"We can give you our running lights," he responded, "but you'll be a sitting duck if you come up over us now."

"Roger that, River Rat. How bad are your wounded?"

"Pretty bad, Dustoff. I'm not sure a couple of these guys are going to make it."

"Let's do our best to see that they do," I told him. "Wait until you hear us approaching from over the river, then give us your running lights.

We'll come overhead with the hoist. Put the two most severely wounded on the jungle penetrator first and strap them on tight."

"Roger, Dustoff, and thank you."

"Don't mention it."

This wasn't going to be easy any way we tried it. Using the hoist and jungle penetrator required maintaining position directly over the pickup site. That was hard enough to do in the daytime over a stationary position. At night it would be extremely difficult due to the lack of visual references. The fact that the boat was drifting would make it even more difficult.

"I've got it," I said, putting my hands on the controls. For some reason I felt this kind of supernatural calm come over me.

"You've got it," Smith said. I could almost hear him breathe a sigh of relief over the sound of the engine and rotors. Of course it was just my imagination.

As we flew up the river, the crew was preparing for the mission. Thank God for the training that afternoon. Pierce and Toomey knew exactly what to do. They had already switched their microphones to the "hot mike" position and were talking to me constantly. I was flying and looking outside, trying to keep my eyes accustomed to the dark and to locate visual references that would help me maintain a steady hover.

I could feel the guys in back getting down on the floor. They were scanning the area below, had found the river and were guiding me toward the boat. The boat captain had turned on his running lights which exposed him to more fire from the river bank. We were totally lights off outside the aircraft, and had only very dim, red instrument lights on inside.

I briefed Smith, reminding him that it was his job to monitor the gauges and talk to me along with the crew. I flipped up the protective cover over the button that would allow me to cut the hoist cable in the event it became entangled. I could control the up and down of the hoist from a switch on the cyclic, but elected to let Tom do it from the pigtail control he had in the back, since he could see better than I could what was happening outside the aircraft.

"Dustoff, this is River Rat three. I hear you coming up the river near our position, but I don't see you."

That's all right, three, we've got you in sight. Hold your position as best you can."

We really didn't need the lights to find them. Their position was obvious by the tracers coming and going in both directions. They were really pinned down so they couldn't get to either bank until they got further downstream. Fortunately, the river was pretty wide at this point. I'd say it was fifty to sixty yards across, and the boat was closer to the bank opposite where the AK tracers were coming from. That would give us some relief.

I approached the boat slowly from the rear, trying to maintain about fifty feet above them. I knew from my training earlier in the day that if I got any lower the rotor wash would make it difficult to load the patients on the hoist. If I got higher, the drift differential between the two craft would make it even more difficult. I was sweating, but didn't have time to wipe the sweat out of my eyes. I had slowed almost to a stop, but was behind the boat. I eased the cyclic forward just a little, then brought it back. I didn't want to overshoot the boat and have to back up.

The crew continued their constant stream of instructions. "A little forward, sir. Right there, now. To your right. Stop. Hold it. Now forward, right there, hold it steady. Hoist going down, sir. Ten feet, twenty feet, twenty-five feet, forward, forward, hold it." The raising and lowering of the hoist had to be done slowly to keep from burning up the motor.

It seemed to take forever. Meanwhile, the boat was moving and I was fighting the wind. "Forward. To your left. Down just a little. Hoist on the deck. Forward. Stop. One patient on. To your right, to your right. Second patient on. Bringing it up."

The crew kept up their chatter. I was concentrating on the horizon, trying not to look at anything too close. The tendency to over-control was great. I had to work at holding the controls very still and making only very slight control movements when they were needed. Thinking about making a control input was almost all that was needed.

We had forgotten about the enemy firing at us. All of us had. We were too busy, and though tracers were flying past, none of them hit us. The crew was still talking. "Forty feet. Thirty feet. Twenty feet. Ten feet. Patients are on board, sir. Let's go!"

I started climbing, glad to get out of there. I was ringing wet with sweat and had a knot in the pit of my stomach. The actual mission may have taken less than five minutes, but it had seemed like an hour or more to me. Toomey reported that both patients were still alive, and he went right to work on them. He told me we needed to get them to the hospital fast. I offered the controls to Smith and pointed him toward Long Binh.

"River Rat three, this is Dustoff seven-four. My medic tells me we need to get these guys to the hospital fast. Can the rest of your wounded wait until we get back?"

"We can take care of the rest of these guys, Dustoff. You take care of those two for us. They're good men."

"We'll be back, River Rat. But it will be about an hour."

We headed for the 24th Evac, alerting the hospital by radio that we were inbound. Thirty minutes later we had unloaded the two men and flew over to Plantation to refuel. Then we headed back to the river.

When we arrived, we found that the patrol boat had managed to bank on the opposite side of the river from its ambushers. They marked a clearing for us to land in and we picked up the remaining wounded men. We offered to take the entire crew in, but the skipper and his men wanted to stay with the boat. I didn't envy them, but they said help was on its way.

We took the remaining wounded men to the 24th Evac, then called it quits for the night. I was so tense when we landed back at Long Binh Dustoff that I just sat in the helicopter to unwind. That had been quite a mission. And it was my first mission since returning from R & R back in the States.

A quiet apology came from the back as we sat in the helicopter waiting for the rotor blades to coast to a stop. "Sorry, I didn't call out the incoming back there, sir." It was Toomey.

"There was incoming?" I joked.

Sandy Pilot Down

Date: 6/23/72
A/C Type: UH-1H
Registration #: 66-15953
Route of Flight: Lai Khe – Local

We were parked in the dust at the edge of the Lai Khe strip when a pair of Phantoms started napalming the area right across the road. Somebody ran over and told us that NVA tanks were less than a mile north of our position and that we had better get airborne. We jumped in, got one burning, two turning, three in the green, and were ready to go. We lifted off and headed south. I contacted the FAC that we'd been working with earlier in the day to find out where the action was so we could stay clear.

"Covey three, this is Dustoff seven-four," I radioed.

"Go ahead, Dustoff seven-four. This is Covey three," he replied.

"We just lifted off Lai Khe and need a sitrep," I told him.

"Roger, Dustoff seven-four," the FAC replied. "Stay three or four klicks south and east of Lai Khe and you'll be clear. Don't get too far away, though. I may need you."

"Roger," I replied. We continued heading south, still climbing. I wanted some altitude. So far there hadn't been any SAM activity south of Lai Khe.

"Dustoff seven-four, what's your fuel status?" Covey three asked.

"An hour and a half," I told him.

When we were above three thousand feet, we turned back toward the east. I saw two flights of Phantoms making bombing runs on a hillside out my left window. Below us a couple of Sandies were circling, waiting their turn. We held up there a few minutes, listening on the airborne C&C frequency to see if we would be needed.

Watching, I was mesmerized by the awesome amount of firepower being displayed right across the road from the airfield we'd been using as a staging area for several weeks. Suddenly I realized the FAC was calling us.

"Dustoff seven-four, this is Covey three, over." His voice was slightly elevated. Something was happening.

"Go ahead, Covey three, this is Dustoff seven-four," I answered him.

"I've got a Sandy down," he told me. Somehow I had missed that. I immediately lowered collective and turned west, scanning the area ahead of us.

"Where is he?"

"I'll mark it for you in just a minute. I don't see any movement down there yet."

I caught site of the O-2, flying west over a small road. "I've got you in sight, Covey three," I told him. "Give us a mark whenever you're ready."

"Roger, Dustoff. I'll be marking with Willy Pete. Look fifty yards south of my mark."

We watched the ground for signs of the white phosphorous rocket that the forward air controller would be firing from under the wing of his O-2. He would be good. Willy Pete is highly flammable and fifty yards was about as close as he would dare put it to friendlies. But it would show well against the green trees.

When he fired the rocket, we immediately saw where it impacted, and I headed for the white smoke, still descending.

"Watch it, Dustoff," Covey cautioned us. "There are bad guys all over that hillside."

"Okay, keep us advised," I told him. "What's the best way in?"

"You're doing fine," he answered. He was watching from his perch overhead. If somebody started firing at us, he would mark the target for one of the Phantoms to take out. I only hoped it wouldn't be too close. The Phantom pilots were typically pretty good and I had seen them drop ordnance within twenty-five to thirty yards of friendlies. But I didn't want to be that close.

"A hundred yards ahead," Covey coached us. We were down on the deck now and he had a much better perspective from his altitude. Suddenly we saw the wreckage in front of us, but we weren't sure how we would get to it. It was in the trees.

I hovered closer to take a look. Then our medic shouted, "Look over there, sir! At two o'clock, there he is!" I couldn't see him, but David Garcia, who was flying right seat, saw him.

"I've got the controls," David said.

I didn't argue with him. He was relatively new in country on this tour, but he was a CW2 on his second tour, and knew what he was doing. "You've got it," I said.

David slid the Huey sideways and started hovering down among a stand of trees. "Keep us covered, Covey," I radioed. "We're sitting ducks down here."

"Roger, Dustoff," he replied. "There's a group of twenty or so bad guys headed your way from the northwest, but I don't think they'll get to you. I've got a little present on the way for them."

"Okay," I told him. "We've located the pilot. Looks like he's okay."

The Sandy pilot, a young Air Force captain, was on his feet with his forty-five in his hand. He was waving us down, but keeping an eye out to the north. Every now and then he would raise his pistol and fire through the trees at a target we couldn't see. We had to take our time hovering down through the trees in order to avoid a rotor strike. Garcia was doing a good job, but I was edgy. I was the AC and we were in a tight place. I was sitting well forward, my hands near the controls.

When we got within five or six feet of the ground, the Sandy pilot jumped on board, grinning from ear to ear. The fuselage of the big old

A-1 was still intact. It had done a good job of protecting the pilot during the crash.

We came up out of the hole in the trees, and followed the FAC's suggestion to depart to the southeast, remaining low-level. He released the Phantoms and they obliterated the A-1, also dropping a load on the North Vietnamese troops that had been trying to reach the downed pilot before we did. We had one happy and grateful passenger on board. He'd been on the ground less than ten minutes when we picked him up.

Last Mission

Date: 9/20/72
A/C Type: UH-1H
Registration #: 69-15223
Route of Flight: Long Binh – YS365480

I must have been touched in the head. It was my last day on the
duty roster, the LZ was hot, and I was letting the Peter Pilot fly. There
were Australians in trouble—Free World Military Forces. They were in
contact, had four wounded, and we were on our way from Long Binh to
pick them up.

It started for us when we got a call in the middle of the afternoon.
My crew was Second-Up and I was hoping we wouldn't have to fly.
First-Up was out around Tay Ninh somewhere on an urgent medevac.
This mission was called in as urgent, too, so it was ours to do.

Smitty did great. He was to become an AC shortly after I was
gone, so I was letting him do the flying. He began picking his way
around afternoon thundershowers east of Long Binh, while I was on the
radio trying to find us some guns. It wasn't going to happen. This was
late '72 and there just weren't many American gunships left in the AO.

Arriving on station we found a classic medevac situation. It was a
hover-hole, the troops were in contact, and us with no guns. It was not
exactly what I had envisioned for my last mission in country.

The Aussies had used some explosives to clear what they called a
landing area. At night it wouldn't have looked so bad, but this was in
the middle of the afternoon, and I could see how small it was. I could
also see that there was shooting going on down there. According to the
smooth-talking Aussie on the radio, they were "taking a little fire from
the north and east." It looked more like heavy fire to me. Smitty made
one of those tactical approaches we had taught him, zeroing out the
Huey's airspeed, then dropping the nose so we would fall like a rock to

the treetops. He headed right to the LZ and honked back on the cyclic only to find it was too small to make a normal landing. We had to come to a hover over it and ease our way down into the clearing. That left us sitting ducks.

We started drawing some small arms fire, but because of the trees the bad guys couldn't get a clear shot at us. Smitty managed to get the helicopter on the ground without hitting any of the surrounding trees, and without us taking any hits. As the wounded were being loaded, the intensity of the firing increased and tracers started coming from everywhere. I could see them impacting the ground directly in front of us and off to our right front and couldn't understand why the bad guys didn't just raise their sites a little and let us have it. It must have been some more of the divine protection that had kept me in one piece throughout my tour.

Being the AC, and being about as short as a body could get, and knowing that if you were going to get killed, your last mission was the most likely time for it to happen, I took the controls. There was no room in the clearing to turn around and the only way out was straight up. I decided to try something drastic. It was something I'd been told about, but had never tried. Something the older AC's had called a "tactical departure."

Telling the crew to hold on, I leveled the rotor with the cyclic, pulled 40 pounds of torque and as soon as we cleared the trees slammed the left pedal against the stop. We went up like a corkscrew and were through 2,000 feet before we knew it. We must have presented a pretty difficult target, because nobody hit us, and there were plenty of people trying. I was also pretty dizzy.

The patients were evacuated successfully and my tour in Vietnam ended. On September 27th, 1972, I trimmed the handlebars off my mustache, passed the urine test at the Pee House of the August Moon at Tan San Nhut Airport, and boarded the Freedom Bird for home.

Chapter 3 – Fort Bragg

My next assignment was to Fort Bragg, North Carolina. The three and a half years spent at Fort Bragg were defining years in my life. During this time I met and married my life's soul mate, my best friend and the mother of my children.

The military provided a number of opportunities for me while I was stationed at Fort Bragg. The first thing on the agenda was to get an instrument rating. A year later I attended school and became qualified as an Aircraft Maintenance Officer and Maintenance Test Pilot, then a few months later, I became a UH-1 Instructor Pilot, and finally a Nap-of-the-Earth Instructor Pilot. I flew medical evacuation missions and functioned as the unit IP for the same medevac unit I had been assigned to in Vietnam.

On the civilian side, I obtained ratings as a commercial pilot qualified to fly instruments in both single-engine and multi-engine aircraft and I became an FAA Certified Flight Instructor. The stories that follow in this section are from my time at Fort Bragg from the end of 1972 through the summer of 1975.

Gliders

Date: 11/11/72
A/C Type: Schweitzer 2-22
Registration #: N9920J
Route of Flight: Chester, SC – Local

I checked in, drew my flight gear, took my standardization check ride, and was ready to fly. My first flight with the new unit was not exactly a tactical mission. One of the other pilots, Mike Novosel, Jr., needed somebody to fly with him to Chester, South Carolina, on a Saturday. There was a glider school there and he wanted to learn to fly gliders. Sounded like fun to me, so off we went.

We preflighted the Huey and made the hour and twenty minute flight to Chester early on the morning of November 11, 1972. I had already decided that adding a glider rating to my certificate would be fun and the cost was reasonable, so when we got to Chester, we both met with our instructors and got with the program.

Learning to fly gliders was a lot easier than I thought it would be. The first flight was a dual flight in which the instructor and I were towed to 3000 feet. On the way up he showed me how to stay behind and below the tow plane so as not to jerk the tow pilot around the sky, and what to do if the towrope were to break or get entangled. Once released from the tow plane, we did some turns, some stalls, and went looking for some thermals. I was surprised at how easily we stayed aloft, even in this relatively low-performance sailplane. Using the spoilers to dissipate lift, we finally brought the sailplane down and landed.

The initial flight was followed by two more tows to 1,000 feet, followed by pattern entry and landing. Then we did a simulated towrope break on the next climbout, followed by a landing.

The next flight was a solo flight to 3000 feet. With just me in the glider, it seemed to have the ability to stay aloft indefinitely. Of course it didn't really, and I had to keep a close eye on the airport to insure that I didn't drift so far away that I wouldn't be able to make it back to the airport when I ran out of lift. It was a nice flight of a half an hour or so and I hated for it to end.

All the time I had been flying in 20 Juliet, Mike had been flying with another instructor in a similar Schweitzer 2-22. By early afternoon, we were both Glider pilots. We cranked up the Huey and headed back to Fort Bragg.

The M.A.S.T Mission

M.A.S.T. stands for Military Assistance to Safety and Traffic. Since military pilots have to maintain their proficiency during peacetime as well as during war, what better way than with an actual lifesaving mission to perform. This was a time before civilian air ambulance usage had become widespread, so we could actually save a few lives, and we weren't competing with civilian operators. It was a great concept and one that would have only worked for the time in which it was conceived and set up.

At Fort Bragg, the 57th Medical Detachment had multiple missions. We were attached to the 82nd Airborne Division, which required that we maintain a high state of readiness, ready to be deployed anywhere in the world within twenty-four hours. Mission Number One was to support the 82nd during its training missions, which consisted primarily of parachute jumps on the six drop zones that were part of the Fort Bragg reservation. We hauled a lot of guys with sprained ankles, broken arms, scratched-up faces, etc., as the result of missing the drop zone or a bad landing.

Mission Number Two was the M.A.S.T mission, which involved transporting civilian patients or military dependents with serious injuries or illnesses from remote spots, highways and from small clinics and hospitals around the Carolinas to larger hospitals where more treatment options were available. We transported victims of traffic accidents, gunshot wounds, and the like to the hospitals at Duke University, the University of North Carolina at Chapel Hill, Baptist Hospital at Winston-Salem, and Charlotte Regional. We transported dozens of premature infants from the smaller towns to the big University hospitals at Chapel Hill and Duke where their survival rate was much higher. For this purpose we used isolettes that plugged into the helicopter's AC electrical system to provide a warm and oxygen-rich environment for the little babies.

A third aspect of our mission was to transport military patients with traumatic injuries, primarily head and back, to the Navy hospital in Portsmouth, Virginia. The duty roster operated just as it had in Vietnam. Each day there was a First-Up, Second-Up and Third-Up crew on duty. First-Up took the urgent evacuation missions, while Second-Up handled patient transfers. Third-Up was there to pick up the slack whenever the first two crews were committed and to do training when time permitted. There were plenty of opportunities for logging night and instrument hours.

Our Hueys were equipped with extended range fuel bladders that provided comfortable IFR reserves for most of the missions we were called upon to perform. We had civilian radios that were used to talk with the various highway patrol and sheriff's departments. And we had something that very few other Army helicopters had at the time— glideslope receivers. We could actually fly full Instrument Landing System (ILS) approaches, at least in some of our Hueys. One or two of them didn't have the glideslope receivers. Naturally, those were the ones we would be flying when the need for flying an ILS approach to minimums came along.

Sunday Morning on the Haw River

Date: 2/6/73
A/C Type: UH-1H
Registration #: 70-16249
Route of Flight: Simmons AAF – Haw River

First-Up duty on weekends was usually pretty quiet. On Saturday and Sunday mornings, the First-Up and Second-Up crews were required to report to the airfield at 7:00 a.m. to perform a preflight inspection and run-up on their aircraft. If there were no missions scheduled, the crews could return home as long as they could be back at the field within 30 minutes should there be a mission.

I had just settled back into bed at home for some quality sleep time, when the Operations dispatcher called with the news that we had a mission. I didn't ask for details over the phone, but instead donned my flight suit and boots and headed back to the airfield.

When I arrived at the helipad, CW2 Jim Thomas, the aircraft commander for the day, was telling the crew chief to mount an internal rescue hoist in the Huey. He explained that our mission that morning was to rescue a couple of guys who where stranded in some trees in a flooded river near Pittsboro, after having been dumped out of a canoe.

We were to fly northwest to the town of Pittsboro, then fly up Highway 501 to the Haw River and turn left. A mile or so up the river we would find a bunch of guys and several vehicles in a field on the south side of the river.

It wasn't raining when we left Simmons Army Airfield, but there were dark clouds overhead and patchy cumulus clouds close to the ground. They were the kind of clouds we pilots like to call "scud" and flying around them was known as "scud-running." We found Pittsboro, flew north up the highway to the river, turned left and within a few minutes saw some jeeps and pickup trucks in a field.

We landed and a Major Majors from the North Carolina National Guard approached the helicopter. Our medic loaned his flight helmet to the Major so he could plug into the aircraft's intercom system and give us the scoop.

The Haw River was flooded from recent rains and was flowing pretty heavily. Apparently our victims had either been caught on the river when it flooded, or had underestimated the strength of its current. At any rate, one man had already been swept downstream and was missing. Two men were visible hanging on the lower branches of some trees, close to where the river bank would have normally been, but far out into water now due to the swollen river. The Major had tears in his eyes as he told us how they had been out there all night and he and his men had tried numerous attempts to get to the two stranded men, but the current of the river was just too strong. They thought they had run out of options until somebody thought to call the Army at Fort Bragg.

This was the type of mission we had trained for often. In fact, my most recent hoist training had been only a few days earlier. We told the Major we would get the men safely on solid ground.

Jim wanted to do the flying which was fine with me. Though we had done practice missions recently, the last actual hoist mission I had done had been in Vietnam over the Navy riverboat. That had been months earlier. We lifted off and Jim hovered to a spot directly over the trees in which the two men were stranded.

Honestly, those trees were at least 75 feet tall, maybe more. It was a long way down through them, but Jim pointed the nose of the helicopter into the wind and held it steady, with the crew in back talking to him all the time. The jungle penetrator is a torpedo looking object with three prongs on it that are folded up so the penetrator can be lowered through the trees. For lifting patients out of the trees, the prongs are folded out horizontally and become seats for the patients to sit on while being hoisted back up. We had doubts that the men down below would know how to get on it, or even have the strength to climb on, so our crew chief, Gaylen Smith, volunteered to ride the JP down to help the men aboard. Since both men were obviously physically exhausted and weakened due to hypothermia, Jim agreed to let Gaylen ride the JP down to get them. I monitored the instruments and kept an eye on our position, while the medic kept up a steady stream of chatter to keep us informed of the progress.

Because of weight limitations, Gaylen would not be able to ride back up with the two men. That made the mission even more treacherous as Jim would be severely taxed in keeping the helicopter steady as it was buffeted by the wind and the JP with men aboard banged its way down and up through the trees two times.

I kept my hands near the controls and watched Jim for any sign that he'd had enough. He kept the Huey rock solid while the two men were brought up and swung aboard. The medic lowered the hoist again and minutes stretched into eternity as it was lowered for Gaylen.

Even though the temperature was cold, Jim was covered with sweat. He told me to get on the controls with him. When the medic indicated that Gaylen was ready to be hoisted back up, Jim relinquished the controls to me. Now it was my turn to hold the Huey steady.

You have to psych yourself up for something like this, yet at the same time you have to relax on the controls to avoid over-controlling. I held the Huey as steady as I could, listening to the medic and Jim's instructions, to "move left, now forward, hold it, hold, back a little." Finally Gaylen was on board and I slid the Huey sideways over to the field and set it down. Our patients hugged us before being led away to a waiting ground ambulance. Major Majors reached through Jim's window and embraced him with a warm "Thank You." He was visibly moved. I wondered if he was chiding himself for not having called us sooner. We made it look like a piece of cake, though actually it had been a very tense mission.

Mountain Retreat

Date: 2/14/73
A/C Type: UH-1H
Registration #: 70-16247
Route of Flight: Mosby Army Heliport – Fulton County Airport, Atlanta

There is a US Army Rangers training camp in the Georgia mountains near Dahlonega. During the latter part of February, 1973, CW2 Al Bazzare and I were sent there with our crew for two weeks to provide medevac coverage while some of the aviation units from Fort Bragg participated in a mountain flying training exercise.

Al and I played a lot of chess during those two weeks, but we also did some flying. We weren't called upon for any medevac missions, but we flew some parachute jumpers one day, and we did our share of poking around the mountains landing on pinnacles and in confined areas just for the heck of it.

One night the whole crew decided that we would go to Atlanta for dinner. Before leaving the training camp, Al and I discussed the fact that it would be dark when we returned and that Mosby Army Heliport was not lighted. No sweat, we decided, especially for a couple of Vietnam Army helicopter pilots that had each done our share of night landings to jungle clearings and rice paddies. After all, this was at least an improved airfield, and even though it was in mountainous terrain, there was a security light on a building at the edge of the airfield that we could use for reference.

While climbing out, we circled over the airfield and tuned in two nearby VORs to get a fix. We then knew that we could locate the grass airfield by flying to a point where two radials from these VORs crossed. We also noted the alignment of the airfield and observed that unless there was a radical wind shift during the time we were gone, we would be landing to the north upon our return. Then it was off to Atlanta for some R & R.

We had fun in Atlanta. It was my first time to visit Atlanta Underground. It was quite late by the time we were in the air over where we expected to find Mosby Army Airfield—sometime after midnight, and very dark. No moon, no stars, just dark.

"Al, do you remember how high these mountains are?" I asked. That seems like a basic fact that you would have thought we would have noted.

"No, I don't," he replied.

"You know when we did this in Vietnam, we always had somebody to talk to on the ground," I reminded him.

"No sweat," he responded. "Let's just line up the VOR radials, fly south from there for a couple of minutes, then turn around and head north. We'll see it."

"Yeah, we will," I said. I really wanted to believe.

We stayed at 5,000 feet until we fixed our position, knowing that there weren't any mountains taller than 4,000 feet anywhere around. At least we were pretty sure there weren't any five or six thousand foot peaks in Georgia.

When the radials crossed, four sets of eyes were looking down. I saw a single light in the darkness below us. It could have been the light that was on the building beside the runway. There were no others. I wondered how the entire camp could be asleep, even if it was after midnight.

We turned south, discussing our strategy. I was flying and Al was busy tuning VORs and adjusting the OBS to make sure we had the right radials lined up. When we had flown south for two minutes, I made a tear drop turn and lined us up with where I thought the runway should be. We both thought it would be a good idea to stay high until we saw the light, then descend at a steep enough angle to keep it in sight. As long as we could see the light there wouldn't be any mountains or trees between us and it. If it was the right light, we should be able to pick out the grass runway with our landing light when we got low enough. If that wasn't it, surely we would know in time to safely climb back to altitude.

We arrived at what I thought was about the right sight picture for a steep approach and I lowered the collective to start us down. I had already slowed the airspeed to around 60 knots and as we descended, I slowed even more. The light stayed steady, but we were both ready to get out of there at the slightest flicker.

Down we came and we could almost feel mountains rising on either side of us, though we couldn't see them. No one was talking. I'm not sure we were even breathing. The crew opened their side doors and peered out into the darkness. They were looking for trees, rocks, anything to indicate we were about to hit something.

I turned on the landing light. Nothing. We kept descending into the abyss.

"This is it," Al said, finally.

"It is," I agreed. The VORs wouldn't lie.

Then we saw it—grass, straight ahead, with trees along each side. Somebody recognized the building, and over to our left, passing beneath us, the refueling area; we were home.

No sweat, routine, piece-of-cake. I terminated the landing in a three-foot hover and air-taxied over to our landing spot. We were Dustoff. This was everyday stuff for us. (*Okay everybody, you can breathe now!*)

Becoming an Instrument Pilot

Date: 4/23/73
A/C Type: UH-1H
Registration #: 70-16247
Route of Flight: Fort Bragg – Goldsboro – Portsmouth – Fort Bragg

Soon after arriving at Fort Bragg I was sent to school to get an instrument rating. You're probably wondering how an Army Aviator who has been out of flight school for more than a year and has already had a tour of duty in a combat zone could not have an instrument rating, but that's how the Army did things back in those days.

We did some instrument training in flight school—sixty hours under the hood in a helicopter and another fifteen or twenty hours in "Blue Canoes." These were WWII vintage Link trainers. The sixty hours of helicopter instrument time were done in TH-13Ts, the military version of the Bell 47 "Whirlybird." You know the one. The front looks like a big bubble and the tail looks like it was cut from a radio tower. You've seen it on "MASH."

The result of all of this flight school instrument training was what the Army called a *Tactical* Instrument ticket. It was supposed to get us out of trouble if we inadvertently wandered into the clouds, but it didn't authorize us to punch into clouds on purpose unless it was absolutely necessary to perform a combat mission.

During November and December 1972, I went through the Army's Instrument Training Course at Fort Bragg and became instrument rated in helicopters. That rating came in handy during the three years I flew medical evacuation missions at Fort Bragg, because we flew those air ambulances in all types of weather.

An instrument rating authorizes a pilot to fly when he or she can't see the ground. In theory, that means you can file an instrument flight plan, get a clearance from the FAA, take off and go right into the clouds, fly across the country without seeing anything outside the airplane but clouds and associated weather, arrive at your destination, and fly an instrument approach to within 200 feet of the ground. Hopefully by then you see the ground or at least the runway lights, and land. In theory.

When the flight that is the subject of this particular story took place I was instrument rated in both airplanes and helicopters, but lacked experience. Without experience, you're likely to botch up, or at least scare yourself pretty badly, if you tackle really tough weather. It's best to tackle weather a little bit at a time, and under the watchful eye of somebody who has a lot of experience. That way you can slowly get your feet wet. Sometimes experience comes to you gradually, and sometimes all at once. I got a bunch of it all at once.

On this particular flight we picked up a patient in Goldsboro that we transported to Portsmouth Navy Hospital in Virginia. Then we flew over to Navy Norfolk to refuel. Weather was moving in and we decided to beat it out, justifying our decision by the fact that the weather was moving to the north and we would be flying south. The other pilot on the trip was Captain Roy Leatherberry, my commanding officer.

We filed IFR, because we had to in order to depart the Navy base. The Navy didn't authorize special VFR flights and didn't have any special rules for helicopters that would let us depart VFR when the field was reporting IFR conditions. We took off and departure control assigned us a northerly heading. I reminded the controller that our destination was to the south and he informed me that I must fly the clearance he had assigned as I wasn't the only traffic they were dealing with.

Leatherberry was flying and that was fine with me, since he was in the right seat and that's where all of the good instruments are in a Huey. We were assigned four thousand feet, which took us into some pretty wet and gray cumulus clouds. When it started getting rough, Leatherberry said to me, "you've got it."

I've got it?!! What did he mean, I've got it? What he meant was, *You're the warrant officer, you're the one who loves this stuff, who signed up to be a professional aviator. I'm a career officer who just flies helicopters because it looks good on my profile when it comes time for a promotion, and I don't like this, and besides, I'm your commanding officer, so "you've got it."*

I took the controls and started trying to catch up to and gain control of the helicopter using the instruments on my side of the cockpit. It wasn't working. A couple of minutes of that and I was ready to quit. But I couldn't quit. Leatherberry had slid his seat back, folded his arms, and was in the *thank God, I'm not flying in this stuff* mode. Meanwhile *"this stuff"* was getting meaner. It was tossing us around like salad ingredients in a mixing bowl.

Something had to happen or we were going to be titanium and aluminum scattered on the ground. I'd like to say that I thought about the crew in back and about how much it would cost the taxpayers to replace the Huey, but the fact is I thought about saving my own butt. I scooted my chair back and scrunched down in the seat and instead of trying to use the pitiful group of instruments on my side of the cockpit, concentrated my scan on Leatherberry's instruments. I had to make some allowances for the angle I was viewing them from and it did put a strain on my neck, but by golly, the scan started working.

When you're doing stuff like this, the only way to make it work is to constantly talk to yourself. You have to keep telling yourself what you are seeing and what you have to do about it. You also have to keep up a pretty good pep talk to convince yourself that you can handle it. I kept it under my breath so the CO wouldn't know how scared I was and how under-qualified I felt, but you'd better believe I was talking to myself, big time.

After a few minutes of this I began to believe the things I was telling myself. I know it was the voice of some of my instructors talking as well, but I began to believe them, too, and I was actually flying the aircraft in spite of its ups and downs and heaves and ho's, and the longer

I did it, the more I realized I could do it. I willed myself to relax because I knew keeping a light touch on the controls was crucial to not over-controlling. I thought back to my horse training days and remembered that the way to stay on a bucking colt is to sit loose in the saddle and roll with the motion of the horse instead of against it. I did that in the left seat of that Huey that day, my seat as far back as it would go and still let me reach the pedals, and scanning the instruments all the way across the cockpit. I did it in spite of Roy Leatherberry's unwillingness or inability to help me and in spite of the controller's insistence on keeping us on a northbound heading when we should be going south.

Finally, we were turned south, which was a relief on the one hand, but on the other hand meant we would have to fly right back through some of the stuff that had just been kicking us around pretty good. Remember, we didn't have weather radar or stormscopes and stuff like that in our Hueys. We just had our eyeballs and my eyeballs were telling me this was stuff we shouldn't be flying in, but every request for a different clearance was met with "standby" or "unable." I knew that stuff was rough because we had just flown through it, but I also knew we had made it through the first time, so could make it through again, and we did.

Somewhere along the way that day, I went from wondering if I could do it, to knowing I could. I went from insecure to confident, from not knowing to knowing. In short, I became an instrument pilot. I know it now, and I think I must have known it that day long ago. As if to reward me for this new discovery, the cloud we were in inhaled deeply, swished us around in its lungs for a few seconds, then spit us out into the clearest, smoothest, blue-sky air one could imagine. "I've got it," Leatherberry said, sliding his seat forward again.

Civilian Ratings

During the Spring of 1973 I started using my GI benefits to obtain the civilian ratings I would need to be a professional pilot. The first order of business was a multi-engine rating. The training airplane for this was a Twin Comanche and my instructor was Fred Harrell, a very capable and professional aviator who flew for the North Carolina Forestry Service and worked as a flight instructor when he wasn't fighting forest fires.

After the multi-engine rating came an instrument rating for airplanes. Again, Fred Harrell was my instructor and we flew a variety of aircraft for the training, including a Cessna 177RG Cardinal and a Cessna 206. The bulk of the training was done in a Cessna 172 and it was in this aircraft that I took my check ride. The airplane instrument rating was added to my ticket in July 1973.

Crossing the Smoky Mountains VFR

Date: 5/26/73
A/C Type: Cessna 206
Registration #: N50518
Route of Flight: Fayetteville – Oxford – Jackson – Fayetteville

"Don't forget to lean the mixture when you reach cruise altitude. And stay out of clouds. Even though you have a helicopter instrument rating, you're not legal yet to fly an airplane IFR." These were the instructions my civilian flight instructor, Fred Harrell, gave me before sending me off on what could be considered my first real cross country solo trip.

I'd flown some cross country in Vietnam, but not a lot. The only trip of any real length had been flown with a much more experienced pilot in the left seat. After returning from Vietnam and being stationed at Fort Bragg, I attended the Army's standard instrument training course to obtain my instrument ticket. Now I was using GI Bill money to get an airplane instrument rating, a multi-engine rating and later a Certified Flight Instructor certificate.

The instrument training syllabus allowed for fifty hours in airplanes, so I was using some of the GI Bill money to visit my family in Mississippi.

The 206 was a big airplane for just me and my German Shepherd, Countess. I'd have preferred making the flight in the flight school's Cessna 177 Cardinal RG, but someone else had booked it for the weekend.

The flight from Fayetteville to Oxford was uneventful. The air was calm and the wind such that I was able to make the flight nonstop. The next morning I took care of some business in Oxford, then flew my sister and her husband to Jackson. The following afternoon I headed back toward Fayetteville, intending to get there before dark. The weather was CAVU at Oxford and forecast to be the same in North Carolina at the time of my arrival. I expected a smooth, uneventful flight.

Trouble started over the Smoky Mountains. It was hazy over the mountains—so hazy that I struggled to maintain visual contact with the ground. Late afternoon cumulus buildups were getting harder to dodge as I made my way along the V-54 airway between Chattanooga, Tennessee, and Greenville, South Carolina. I began encountering strong up and down drafts and had no idea which way to go to avoid them. Towering cumulus clouds were popping up everywhere. Weaving my way between them I was finding it harder and harder to stay clear of clouds. At one point the airplane was descending at 2,000 feet per minute in a strong downdraft. I had the nose pointed up and the engine at full power, but the Cessna was unable to pull itself out of the descent, even at maximum rate of climb airspeed. During the descent I tried maximum rate of climb airspeed and maximum angle airspeed both. Either way, we were still going down. Just when I thought we'd plunge all the way into the ground, the downdraft spit us out—right into the hands of an equally strong updraft. This time we were on an express elevator ride to the heavens.

Countess was amazing. She made her way back to the luggage compartment and rode there without complaint. I promised her a smoother ride and turned south, the shortest way out of the mountains.

An hour later, I was able to relax as I found smoother air, fewer clouds and better visibility. In later years, as I flew all over the United States, I learned that flying over the Smoky Mountains could present some of the most interesting weather challenges a pilot might encounter. In fact throughout my flying career, there have been just a few times when I had to land and spend the night somewhere just to wait out the weather. Most of those have been related to trying to cross the Smoky Mountains in an unpressurized or non-turbocharged airplane.

First Trip to Walter Reed

Date: 9/12/73
A/C Type: UH-1H
Registration #: 70-15706
Route of Flight: Fort Bragg – Walter Reed Army Hospital

On my first trip to Washington, D.C. as an Aircraft Commander, I screwed up. The mission was to transfer a military patient with a back injury from the hospital at Fort Bragg to Walter Reed Army Hospital.

We left early in the morning and enjoyed a smooth flight, though it was a long leg for a helicopter. We had been in the air for over three hours when we were handed off to Washington Approach. The controller verified that our destination was Walter Reed and offered to provide vectors. We flew on for a few minutes and he told us that Walter Reed was at our twelve o'clock position about eight miles. I thanked him and began looking for a hospital helipad.

Very soon after that I saw a big hospital complex and beside it a helipad with a big red cross painted on it. I told the controller we had Walter Reed in sight and thanked him for his help. I lowered the collective lever to initiate an approach to the helipad.

When we landed, a military ambulance drove up and two hospital orderlies got out of it and approached the helicopter. One of them came up to the window on my side of the aircraft.

"What have you got?" he asked, yelling over the sound of the helicopter.

"Patient transfer from Fort Bragg," I told him.

He shook his head. "We're not expecting any patients from Fort Bragg."

"You're not?" I was puzzled. Of course the transfer had been pre-arranged.

"Are you sure you have a patient for Bethesda?" he asked.

"Bethesda? No, we're supposed to be at Walter Reed," I told him.

"Walter Reed is about five miles that way," he pointed.

I thanked him and as soon as he was clear of the aircraft I rolled the throttle back to full rpm and we lifted off. No sooner had we broken ground than the radio came to life.

"Medevac 706, this is Washington Approach. Landed at the wrong hospital, didn't you?"

"Sure did," I admitted.

"I was afraid of that when I saw you disappear off the scope, but couldn't raise you on the radio that low. Walter Reed is now at your twelve o'clock and five miles."

He began to describe the landing area to me in great detail. He told me I'd be landing on a ball field with a red clay cliff at the south end and a bunch of temporary buildings at the north end. As he described the landing site, it came into view. It was obvious from his description that we were now at the right place. In fact the ground ambulance was already there waiting for us.

What a contrast! Bethesda Naval Hospital had been all white and clean and modern looking. Walter Reed looked like a series of World War II Army Barracks and pretty run down at that. I should have known, since I'd been fortunate enough to live with the Navy during the first half of my Vietnam tour, that the Navy goes first class, while the Army makes do with whatever it has.

First UFO Sighting

Date: 10/19/73
A/C Type: Cessna 177RG
Registration #: N1892Q
Route of Flight: Fayetteville, NC – Oxford, MS

I have a couple of UFO stories. I guess being in the air a lot gives you opportunity to see things you wouldn't otherwise see. I'm not saying these are extraterrestrial sightings, but they're very interesting, nevertheless. This is the first one.

Joyce and I had been married almost three months. Since I had a little GI Bill training money left that I could use for flying, I talked her into going to Oxford for the weekend. Joyce, Countess and I left after work on Friday and planned to be back in Fayetteville Sunday night.

The Cessna Cardinal 177RG is a very comfortable airplane and with its retractable gear, it has a decent cruising speed. Making it to Oxford nonstop was no problem, even flying west into the ever-present headwind. We were on an IFR flight plan at night, cruising at 8,000 feet over Chattanooga when we saw a green glow in the sky above us. You know how the sun reflects off clouds at dusk and they glow sort of orange or maybe even red? It was like that, only it was after 11:00 at night and the glow wasn't red or orange. It was green.

No, it wasn't swamp gas. There are no swamps around Chattanooga. Plus, we were more than a mile and a half up and this green, glowing cloud thing was well above us.

We weren't the only ones that saw it. There was a lot of chatter about it on the radio. Several airline pilots commented about it to the controllers and to each other. There were some other general aviation pilots flying through the area and they all saw it, too. Funny thing was, not a one of us could offer an explanation for what we were seeing, not even speculation.

The airline pilots told the rest of us that as they climbed they finally got above it, but it seemed just as strange from above as it did from below. They estimated its altitude to be between 12,000 and 15,000 feet.

We flew on toward Mississippi without getting an explanation for what we saw and to this day don't have a clue.

Cross Country Challenges

Date: 3/22/74
A/C Type: Cessna 172
Registration #: N20058
Route of Flight: Fayetteville, NC – Columbus, GA

Joyce's brother was getting married in Columbus, Georgia. We decided to fly down for the wedding in a Cessna 172. Joyce's sister, Jeannie, and her husband, Den, accompanied us as passengers.

The flight was IFR, both coming and going. We weren't in the clouds all of the time, but I did get at least three hours of actual IFR along the way. There are two things I remember about the flight that are part of the instrument cross country learning experience. One of the events happened on the way down, the other on the way back.

On the way to Columbus we were cruising along at 6,000 feet, in the clouds, but in smooth air, men up front, women in back. We were talking and enjoying ourselves, when the engine began to surge and lose power. I went though my power loss checklist. The first item on the list, apply carburetor heat, solved the problem and the engine smoothed out and behaved itself for the rest of the flight.

On the way back, we were cruising along at 5,000 feet, again in the clouds when the VOR I was navigating to went off the air. I tried the other radio and determined that the problem was on the ground; it was not a problem with the equipment in the airplane. I called the air traffic controller.

"Atlanta Center, this is Cessna two-zero-zero-five-eight. We're no longer picking up the Florence VOR."

"Roger, Cessna zero-five-eight," he replied. "Florence is down for maintenance." They did that in those days. At least I guess they did. I don't recall it ever happening since.

"Well, my clearance is up the airway to Florence, then a turn toward Fayetteville," I told him. "How will I know when to turn?" Remember, this was in the days before RNAV or Loran, or GPS. VORs were all we had.

"I'll tell you when to turn," he replied. Thank God for radar.

It was simple enough, but in those early days of IFR flying, it caught me off guard. The controller told me when to turn and a little further along I picked up the next VOR. After that the flight proceeded without further incident.

Maintenance Officer

My dad's advice when I went into the Army was to take advantage of all of the training I could get. I thought that was good advice, so when I was offered a chance to attend the Aircraft Maintenance Officer Course at Fort Eustis, Virginia, I jumped at it.

My advisors told me that I was making a career choice. They explained that as a Warrant Officer Aviator in the Army I would have a choice of one of three career paths.

1. Maintenance Officer/Test Pilot
2. Training/Instructor Pilot
3. Operations/Flight

Either of the first two options would have been all right with me, but my obvious choice would have been number two. But they weren't offering me number two, they were offering me number one. My commanding officer, a Major, explained that if I chose that particular career path, I would be locked in to it. I chose it anyway, because it was an opportunity for training. It turned out that he was wrong. Just a few months after returning from the five-month course at Fort Eustis, the 57th had a slot available for a Huey Instructor Pilot and for some unforeseen reason, I was the only one available to go. I don't know of anyone else that became dual qualified as a Maintenance Officer/Test Pilot and an Instructor Pilot so early in their career (if at all) but I considered it a blessing right out of heaven and took advantage of both schools.

The Maintenance Officer Course involved four months of classroom work in which we learned all about aircraft systems. We studied engines and airframes, electrical systems, hydraulic systems, landing gear systems, environmental control systems, etc. The training didn't just cover the Huey, but all aircraft in the Army inventory, both airplanes and helicopters. As an aircraft maintenance officer, we wouldn't be expected to actually work on aircraft, but we would be expected to manage, schedule and oversee all of the maintenance work. The manuals they issued us literally filled two army duffle bags.

The fifth month in the course consisted of Test Pilot training. Upon completion, students of the course were qualified to perform maintenance test flights on the particular aircraft that they were assigned to maintain. For me that was the Huey.

It was a rare opportunity to learn the Huey inside and out. Each day during the course one or more malfunctions were "programmed" into the aircraft. A safety wire was left loose, a bolt or nut was loosened, a fuel line was disconnected, or something similar.

We were supposed to catch the discrepancy during our preflight inspection, but if we didn't, we were allowed to fly the aircraft with a very capable instructor pilot along. Chances are that if we didn't find the problem on the ground, it would show up in flight. It was then our responsibility to take the appropriate action, including responding to an emergency if such occurred, recover the aircraft to a safe place, diagnose the problem, oversee the repair, then fly the aircraft again to insure that the problem had been properly corrected.

Talk about a confidence builder! At the end of this course each of the participants felt like we could fly a Huey with one rotor blade missing, if necessary.

The school at Fort Eustis ran from late October 1973 through mid-March 1974. After returning to Fort Bragg I was given an instrument check ride, a 90-day standardization ride and put back on the duty roster. The Maintenance Officer slot in our unit was already filled by a very capable individual, CW2 Gary Reed. Gary made sure I got on test flight orders and got to do a few test flights with him, but I never did work full time as Maintenance Officer.

I did learn a lot under Gary's tutelage. One of the things I found fascinating was balancing and tracking rotor blades. On a helicopter the blades have to be balanced, one with the other, and adjusted so that they track evenly with each other or the helicopter will have either a lateral or a vertical vibration. A slight vibration might just be uncomfortable for the crew, but too much vibration could cause the ship to shake itself to pieces.

Blades are balanced at the factory. We could make minor adjustments to them in the field by adding weight in a hollow place just inside the blade tip. Tracking, however, has to be done almost any time there is work done on the rotor system. You can get it close by counting the number of threads showing on the adjusting linkage. Final adjustments are made by bending trim tabs that are located three fourths of the way out on the blades.

These days there are very sophisticated methods of checking the track using strobes or lasers, and the measurements can be taken at various rpm settings and adjustments made so that the rotor system is very smooth. We didn't have any of that equipment in the early seventies, but we did have a pretty good manual technique.

We would use grease pencils to color the tips of the blades. One we colored red, the other we colored black. Then we took a pole wrapped with foam rubber and covered entirely with masking tape and held it up vertically, carefully moving it inwards so that it just barely touched the tips of the rotor blades with the engine running at flight idle. If the marks were pretty close together at this speed, we would put new tape over those marks and signal the pilot to run the rpm up to flight speed and try again. Any time the marks were too far apart, the pilot would shut down the helicopter and the blades would be adjusted with either the pitch adjustment linkage or the trim tabs. The rule was always to "bring the lower blade up" to meet the other one, if possible. Sometimes the linkage wouldn't allow this so you would have to bring the higher blade down slightly. This procedure would be repeated until the rotor blades tracked perfectly and there was no vibration.

Whenever I did the tracking it would normally take several tries to get it close. Gary could usually do it on the first or second try.

Carrier Qualification

Date: 4/18/74
A/C Type: UH-1H
Registration #: 70-16238
Route of Flight: Simmons AAF – USS Guadalcanal

I failed to mention that during my Vietnam tour I became Carrier qualified. Yep, that's right; I was an Army pilot who was Carrier qualified. All of us in the 57th were required to maintain qualifications to land on Navy vessels because the ones nearby offshore were in our area of medevac coverage. Not only did we land on aircraft carriers, but on more than one occasion we flew out offshore and landed on a destroyer to pick up a patient.

Now, back at Fort Bragg, we were once again being tasked with obtaining and maintaining proficiency in landing on Navy ships. In a helicopter it's nothing like what the fighter pilots have to do, but it is an interesting exercise.

The biggest challenge we Army pilots faced was learning the lingo and procedures. We were told to "fly the ball" and the "deck is green" and other phrases that meant nothing to us. When the Navy controllers threw those terms at us, we just pointed our helicopters at the deck and flew our approaches like we knew what we were doing.

Helicopters approach aircraft carriers from a forty-five degree angle off to the side. I guess that's to keep us out of the way of fixed wing traffic. The trick was to make the approach to a point above the deck then plant the helicopter on the deck firmly by lowering collective. The fact that the landing spot is moving sideways as you approach it, doesn't present too much difficulty because you subconsciously adjust to it as you are making the approach. It is the touchdown that is tricky. I learned to time my planting with the rising and falling of the deck. The deck would pitch up—one, then down and up again—two. The third time the deck came up, I'd lower the collective to meet it and we'd be down. Immediately the deck crews would run out with chains to anchor our skids to the rolling deck. It was possible to slide overboard if not secured.

We always took advantage of the opportunity to eat with the Navy when on these carrier missions. I had learned in Vietnam that the Navy has exceptionally good food in peace time and during war.

I never had a problem with seasickness on the deck of an aircraft carrier, but sometimes below, while walking down the narrow hallways I would get a little queasy. I didn't let on though and by the time I was seated at the dinner table I was fine.

Worst M.A.S.T Mission

Date: 5/3/74
A/C Type: UH-1H
Registration #: 70-16238
Route of Flight: Simmons AAF – Dunn – RDU

One of the most frustrating flights of my air ambulance career occurred on the night of May 5, 1974. Around ten or ten-thirty that night we got a call to pick up a gunshot wound victim in Dunn, a small town just north of Fort Bragg. We were to take him to Duke University Hospital.

It was a rainy, overcast night with lots of low-lying clouds around. Fortunately our working relationship with Fayetteville Approach Control and Raleigh Approach Control provided us with a lot of flexibility. Our medevac call sign afforded us priority handling and when we needed a clearance we could get it.

My copilot that night was a WO1 newly out of flight school. Because of his limited experience, most of the burden of the flight would be upon me. I chose to fly from the right seat where the best instruments were located.

We departed Simmons and I began working out the details of the flight with Fayetteville Approach Control. An ambulance in Dunn was going to bring the patient to the grass airstrip at Dunn to meet us. We stayed beneath the clouds and flew northeast toward I-95. Approach Control worked on getting us an IFR clearance out of Dunn for Raleigh-Durham Airport (RDU). The current weather at RDU was 700 feet overcast with 2 miles visibility in fog and light rain. If it held we would be able to get into Duke after flying an instrument approach at RDU. If the weather didn't hold we might have to terminate the flight at the airport.

The patient was a soldier in his early twenties. When they brought him to the helicopter at Dunn, we could see that half his head had been blown off. The story we got was that he had been playing Russian Roulette with a .45 and lost.

The medics were breathing him with an ambu bag. It didn't look good. We broke ground and headed into the clouds. It would have been foolish to try to low-level it all the way to Durham.

En route at 2,000 feet I was getting regular updates from our medic in back about the condition of the patient and from Fayetteville, then Raleigh Approach about the deteriorating weather at RDU. It soon became apparent that we weren't going to get into the soccer field that we normally used for a landing field near the medical center at Duke University. I advised Raleigh Approach that we would need an ambulance to meet us at RDU for a critical patient. He rogered and issued an approach clearance for the ILS to runway five.

Asking the controllers to call an ambulance to meet us at the airport was not unusual. The controllers at Raleigh had done it on several occasions before, as had air traffic controllers in other areas around the Carolinas and Virginia. This part of the mission was fairly routine, though we were concerned that our patient would not survive the twenty minute ambulance ride from the airport to the hospital. But what else could we do?

I still had it in my mind that if we broke out with enough visibility to safely make the hospital, that's where we would go. But as the approach clearance was issued, the weather was given as 200 foot ceiling and one mile visibility—ILS minimums.

Then the kicker. The helicopter we were in didn't have a glideslope receiver installed. That meant a localizer approach which has higher minimums—400 and two. Since we were in a helicopter, I was authorized to commence the approach with as little as a half-mile visibility. Because it was a non-precision approach, the technique was to fly across the outer marker (approximately 4 miles from the end of the runway) at a predetermined altitude, then immediately descend to the minimum descent altitude which was 400 feet above the ground and maintain that altitude until a certain amount of time had passed, or we saw the runway, whichever came first.

I hit the 400 feet, and because we still had some time to go, eased back up 50 feet so I wouldn't bust minimums. When he saw me climb back up, my copilot called the tower and told them we were executing a missed approach. My fault, I should have briefed him on what I was doing. But we weren't executing a missed approach and I immediately got on the intercom and told him, "no, no, no!" I also made the mistake of looking at him, and that moment of taking my eyes off the instruments and turning my head gave me a bad case of vertigo.

I was fighting to recover from this when all of a sudden the world was lit up by bright flashing lights. It was the approach lights, but bouncing off the fog the way they were, they were blinding me.

I keyed the microphone and yelled at the tower, "Turn the approach lights down!" The controller must have anticipated the request because the lights immediately dropped to a much lower level of brightness. When I recovered from my vertigo, we were much lower than we should have been and just above the approach lights. I flew on to the runway.

As we turned off onto the ramp the tower operator told me he had called the ambulance and they should be on their way.

As I was shutting down the medic told me we were losing our patient. He was working the ambu bag and the crew chief had started CPR. The ambulance didn't arrive. We knew our guy was gone, but we weren't authorized to call his death, nor did we want to quit. Heretofore, I had never lost a patient on my helicopter, not even in Vietnam.

We took turns pushing on the guy's chest and breathing for him with the hand pumped ambu bag. It was hard work. Finally, I headed up to the tower to see if I could find out what had happened to the ambulance. As I walked off, the ambulance showed up. No lights, no siren. Two paramedics got out and came over to the helicopter. They looked at our patient and said, "he's a stiff!"

"He wasn't when we called you," I said. Inside I was steaming. "Where have you guys been?"

"Oh, we stopped to get something to eat," one of them told us. "Nobody told us this guy was critical."

I was speechless. I was so mad I couldn't think straight as they drove off with our now deceased patient.

You better believe that when I got back that night and the next day I made phone calls. I ranted and raved, I demanded explanations, but it was all to no avail. The guy was dead. It could have been that he would have died anyway. Who knows? But it was my first patient to lose and I didn't like it.

Instructor Pilot

Because I had attended the Aircraft Maintenance Officer Course a few months earlier, I was completely surprised in early June when my commanding officer congratulated me on being selected to attend the UH-1 Instructor Pilot Course. The truth was that they didn't want to send me, but no one else in the unit was available to go.

Normally the IP school was held at Fort Rucker, Alabama, the home of US Army Aviation, but since there were a number of candidates at Fort Bragg that needed the course, Fort Rucker sent its finest up to help conduct a course on local soil. Local IPs would teach the course and Fort Rucker Standardization IPs would administer the final check rides. My primary instructor for the course was Jim Miles, the 57th's existing IP and a fine gentlemen that I had flown with many times.

The course consisted of ten days of concentrated flying, three or four hours a day, all of it emergency procedures. On one of those days, Jim taught me something I had never seen before. It was called a "back up" autorotation. We had practiced all kinds of engine out procedures including normal autorotations, 180 degree autorotations, low-level autorotations and hovering autorotations. These were all maneuvers I had seen and practiced before, but never with such intensity.

The back up autorotation was something entirely new. Its purpose was to allow you to land at a spot that is directly underneath the aircraft at the time the engine fails. The procedure is to apply aft cyclic to zero out the airspeed, then start the helicopter moving backward. When the desired landing area, that was previously just below you, appears ahead at an angle like that of a normal autorotation, you dump the nose, pick up airspeed and perform a normal autorotation. Pretty slick.

During one of the days of training, Jim and I had a close call, all my fault. We were at Camp McCall, which was an old WW II Army Base southwest of Fort Bragg. Camp McCall had one of those old triangular runway layouts common to military bases of that era. It was used by Fort Bragg personnel for training on numerous occasions. Doing the IP training at Camp McCall kept us away from the traffic at Simmons Army Airfield.

We were doing an anti-torque landing, which simulates loss of tail rotor thrust. The idea is to make a running landing using the slipstream to help keep the aircraft straight and using the throttle to add or reduce the effects of torque on the airframe's alignment. Rotating the throttle counterclockwise causes the nose of the aircraft to turn left and rotating it clockwise causes the nose to turn right. The correct procedure is to approach with the throttle retarded to about 6200 rpm (6600 is normal) which allows the nose to be slightly left of center. Just before touching down you're supposed to roll the throttle on, thus using torque to align the aircraft's nose with the direction of flight. To save the skids, we were landing on the grass beside the runway, rather than the runway itself.

Everything was looking good, but the terrain was a little uneven with high grass and we touched down before I thought we would. I still had the nose to the left, so the right front skid of the Huey dug into the ground and before I knew it, the helicopter was rolling to the right. I thought we were going to turn over.

My mind went into slow motion and began analyzing how to get out of a crashed Huey. What I should have been doing was analyzing how to keep the Huey from crashing and that's exactly what Jim was doing. He flew us out of the "crash."

It was so close to an actual crash that the crash crew had jumped into their truck and started heading toward us. The tower operator told us later that he knew we had crashed and was totally surprised when he saw us fly out of it.

Jim flew us around the pattern and landed beside the runway, where we got out to inspect the damage. The right front skid tube was bent out at a fifteen to twenty degree angle. There were grass stains on the tips of the rotor blades, but fortunately no damage there.

The Huey we were flying was not one of the 57th's, but had been borrowed from the 119th Assault Helicopter Company. The maintenance officer from the 119th came over to inspect the damage. He and Jim had a little discussion about whether or not we would have to report the incident and Jim was concerned about his previously spotless record. The maintenance officer told him not to worry about it, that he would take care of it and apparently he did. He assigned us another helicopter for the day and took that one back to Fort Bragg. Apparently everyone involved with the program was interested in maintaining a perfect safety record for the course and it never was reported. The Huey got a new set of skids and that was that.

I learned something from the incident that has had a profound affect on my flying. Jim said it this way: "Don't ever quit flying the aircraft. As long as you have control, use every bit of control you have." I had resigned myself to crashing, but Jim worked on ways to avoid crashing by using the available rotor rpm, cyclic and pitch control to fly the aircraft to a safe location.

On June 19, 1974, I successfully passed the end-of-course check ride administered by another IP and two days later flew with a CW4 from Fort Rucker for my IP Qualification check ride. He passed me, but said my autorotations were "too smooth" and that he wanted me to fly one more session with my course IP before being turned loose as an Instructor Pilot.

Jim was pretty upset by that "smooth autorotation" diagnosis. Of course my autorotations were smooth. We had worked on making them smooth. I was doing them just as I had been taught with a little added smoothness due to my own superior pilot technique (just kidding). Nevertheless, Jim and I went up for one more session and after that I became the 57th Medical Detachment's new IP.

I loved the role. Teaching comes natural to me and is something that I enjoy doing. This job didn't involve teaching beginners how to fly, but it did give me the responsibility of making sure that each of the pilots in the unit had opportunity to practice emergency procedures on a regular basis under the watchful eye of an instructor pilot.

The next few months I alternated between flying missions, flying maintenance test flights and giving check rides to other pilots. This gave me an excellent opportunity to constantly improve my own piloting skills and I began to wear the Huey like a second skin.

Asking Directions

Date: 10/20/74
A/C Type: UH-1H
Registration #: 70-15706
Route of Flight: Fort Bragg – Winston-Salem, Moses B. Cone

The mission was to take a heart patient to the Moses B. Cone Hospital in Winston-Salem. I was familiar with most of the hospitals in North Carolina and had been to the Baptist Hospital in Greensboro many times. This was my first trip to Moses B. Cone and wouldn't you know it, it was a night mission and a very dark, overcast night at that.

Joe MacDonald, the former IP, and I had created a reference book with photos and information about most of the hospital helipads or landing sites in North Carolina, but this one wasn't in the book. It was a big hospital, however, and both the Winston-Salem police and the hospital knew we were coming. We hoped they would have the place well marked for us.

We arrived over Winston-Salem and began to look for flashing lights. I asked the approach controller for help, but he didn't have the hospital location marked on his scope and wasn't familiar with it. I flew over the city at a fairly low altitude figuring the hospital would be on a major thoroughfare and we would recognize it as a hospital when we saw it.

I felt kind of stupid at not having gotten explicit directions to the hospital before leaving for the mission, but here we were wandering around over the city, looking for something that we had no idea where it was or what it looked like. Finally, the crew chief said, "Sir, I see a police car below us with his emergency lights on and he's shining a spotlight up at us."

I turned in the direction he told me and saw the police car sitting beside the road near a cloverleaf intersection on what looked like an

Interstate highway. It seemed a natural thing to land and see if he could help us, so I did. Landed right there in that cloverleaf.

The cop came running over to the helicopter, stuck his head in my window and asked, "Are you the Army chopper that's looking for the hospital?"

"Yes, sir, we are. Are we anywhere near it?"

"No," he answered, "but I'll take you to it, if you'll follow me." With that he ran back to his car, got in and started off down the highway with his emergency lights flashing.

I looked over at the copilot, John Harris, and told him, "I sure feel silly doing this."

"I know what you mean," he said, and I took off in pursuit.

Every time the police car made a turn, we turned with him, flying about one hundred to two hundred feet up. Finally, he arrived at the hospital and pulled up beside a brand new heliport, which was why we hadn't known about it. An ambulance was waiting and the patient was unloaded.

As we sat there, preparing to depart and go to the Winston-Salem airport for fuel, I looked at John, who was also an experienced combat veteran pilot. "Why didn't we get some altitude and just watch where the guy was going, instead of following him around the city like that? We could have seen the whole city from five or six hundred feet up."

"I don't know, man," John answered. "You were flying."

Long Range Ferry Trip

Date: 12/1/74
A/C Type: UH-1H
Registration #: 70-15706
Route of Flight: Fort Bragg – Charlotte – Greer – Atlanta – Birmingham – Oxford – Jackson – Fort Polk – Lufkin – College Station – San Antonio – Navy Corpus Christi

One of our helicopters, 15706, was very high time and due for a total overhaul. She had a personality all her own and a shuffle about her. It could have been the patches covering all of her bullet holes. Our operations officer, Captain Terry Muldoon, had been shot down in 706 in Vietnam. Apparently he wasn't the only one.

When the time came to ferry 706 to the overhaul facility at Naval Air Station Corpus Christi, the duty roster rolled around in just such a fashion that WO1 Rich Seifried and I were assigned the flight. That was fine with both of us because it meant a lot of flying hours and the possibility of visiting both my family in Mississippi and his family in San Antonio along the way.

We were scheduled to leave on a Saturday morning, but when we got to the airport, the weather was IFR. Army regulations didn't prohibit us from flying IFR en route on a ferry flight, but they did prohibit us from beginning the ferry flight under IFR conditions.

We weren't sure we would be able to get out of Simmons VFR at all that day, but as chance would have it, the clouds lifted enough around mid-morning that we were able to depart under VFR conditions. Following the letter of the law, we took off and headed west. Within fifteen minutes we were talking with Fayetteville Approach control and getting an instrument clearance to Charlotte, our first fuel stop.

On the way to Charlotte we flew through some pretty wet clouds with the temperature right around the freezing mark. Ice was a big concern. We decided that we would remove the turbine inlet covers at Charlotte before beginning the next leg of our journey. This would at least prevent the engine's air supply from being cut off due to ice.

On the ILS approach into Charlotte, the tower operator passed along the message that "Hoot says hello." That was a surprise, but as we learned once we got on the ground, one of our fellow 57th pilots, WO1, Jim "Hoot" Gibson, who was on leave, had a layover at Charlotte on his way back to Fayetteville and had decided to visit the tower. Imagine his surprise while in the tower to hear one of his own unit's helicopters call in over the outer marker on the ILS approach.

While the helicopter was being refueled, we quizzed other pilots that landed at Charlotte about the weather conditions they encountered on their way in. We wanted to know if they had encountered any icing. None had, but most were flying jets or turboprops and had spent very little time in the clouds. Heading west we would likely be in clouds for a while so we were very concerned about ice. We should have been.

The next leg of our flight was to Greer, South Carolina, only forty-five minutes away. At about the halfway point we began picking up ice on the edges of our windshield and on our skids. We were glad we had removed the engine air inlet screens to prevent them from icing over.

We were at four thousand feet and couldn't go lower because of high terrain. All of the weather reports indicated that we would have the same conditions at 6,000 or 8,000 so we remained at four, hoping the ice wouldn't build up too much.

Although the helicopter flew fine, we knew it was carrying a load of ice. We just didn't know how much. I had read that if you start getting ice in a helicopter you should frequently move the pitch control up and down and change the tilt of the rotor system to keep the control rods from icing up. I did that and I prayed. When we landed at Greer, we found one and a half to two inches of ice on all of the components of the rotor system except the blades. We had to chip it off.

The next leg of flight down to Atlanta had better conditions. We were below the clouds the entire way, but when we arrived in Atlanta it was snowing. It was just a light snow, which didn't affect our flying at all, except for reduced visibility. Still, we had more than the mile required for VFR flying in a helicopter, so we motored on.

All the way to Birmingham we were flying in light snow beneath the clouds and with at least two miles visibility. The afternoon wore on and it was beginning to look like we weren't going to arrive in my hometown of Oxford until after dark.

The plan there was not to fly to the airport, but to my grandmother's farm a few miles north of the airport. She had an open pasture across the road from her house and I knew I would have no trouble finding it even in the dark. There were power lines at the edge of the pasture, but I knew where these were, too, and knew that if we approached from the north and landed at least 50 yards from the fence we would be well clear of the power lines.

We arrived over Oxford around 6:00 p.m. and it was snowing there as well. Because of the overcast and the time of the year it was very dark. I found the road leading out of town that I knew very well and followed it to the little community where my grandmother lived. We were flying about 500 feet above the ground as we arrived over her farm. After locating the house I flew up to the north a little bit, then turned and made an approach to the pasture that was across the road, landing directly in front of the house. The landing was uneventful.

Uneventful for us, that is. I had no concept of how much ruckus a Huey would make at night in a small community. For one thing, it was very loud, even sitting on the ground with the engine at flight idle for the required two-minute cooling off period. We had both landing lights on, the one just under the nose and the other under the belly, plus the anti-collision light on the top of the engine compartment behind the main rotor system. When you're sitting on the ground at night with the engine running, you're supposed to put the red, green and white navigation lights in the flashing mode. We did that and were lit up like a Christmas tree. That's why to all of those country people passing by on their way home from work we must have looked and sounded like a UFO.

My grandmother told us when we got inside for dinner that even though she had known we were coming, the sheer awesomeness of it all had nearly given her a heart attack.

The first order of business when I got inside was to call the FAA and close my flight plan. The Flight Service Specialist kept quizzing me, "are you sure you're on the ground at Oxford?" What I didn't know then, but found out the next morning was that they had alerted the airport manager at Oxford that an Army helicopter was inbound and when we didn't land at the airport, he had reported us as overdue. The next morning when we flew over to the airport to get fuel, the airport manager, Jeff Meaders, said, "If I had known it was you, I would have guessed you would be landing at your grandmother's house, but I didn't know it was you, so I thought we had a missing Army helicopter."

They talked about that event for a long time around the little community of College Hill. I had no idea we would stir up so much activity with our little stopover.

The next morning brought clear skies, so we headed on to Texas. Because it was a ferry flight, we didn't have auxiliary fuel tanks installed, limiting our range to two hours. That meant frequent fuel stops, which had to be planned for locations where there was either government fuel or contract fuel, whenever possible. We flew from Oxford to Hawkins Field in Jackson and from there to Fort Polk, Louisiana. The next stop was College Station, Texas, then over to San Antonio, where we spent a night with Rich's parents and enjoyed some excellent Mexican food.

The next day when we flew down to Corpus Christi, we were told to fly low-level along the beach and to call Navy Corpus Christi for clearance prior to crossing the end of their active runway along the beach. We flew just offshore along "Million Dollar Highway," a stretch of land along Corpus Christi Bay lined with beautiful mansions.

Navy Corpus had us hold short before crossing their runway centerline, which basically had us hovering out over the water for a few minutes. Then they cleared us across and gave us vectors to the Army Depot where old 706 would be disassembled and made like new. Later that day we flew home on a commercial airliner.

Foggy Cross Country

Date: 12/17/74
A/C Type: UH-1H
Registration #: 73-21859
Route of Flight: Simmons AAF — Portsmouth Naval Hospital

I was the IP, Paul Cave was a new pilot. I was a Vietnam veteran helicopter pilot. I was supposed to know what I was doing. Paul was having his doubts.

We were going to Portsmouth Navy Hospital, a trip the crews from the 57th made fairly often. It was 7:30 in the morning and we had been flying for half an hour. Paul kept looking over at me and I could tell he was uneasy about something. Everything on the helicopter was working fine. We were flying straight and level in VFR conditions at 3,000 feet. I decided to wait for him to voice his concerns, rather than volunteer information that might set him at ease. Just a little mean streak I guess.

Finally he voiced it. "You're not worried about that fog down there?"

It was beautiful, the fog was, covering the ground like a blanket of snow as far as the eye could see. It had moved in right after we had departed Wommack Army Hospital. I hadn't given any indication to the crew that I had noticed, but of course I had. I just wasn't worried about it. The instructor in me had been dying for him to ask. Now I could explain.

"I'm not worried," I told him. "It's forecast to burn off before we get to Portsmouth."

"Yeah, but what if it doesn't?"

"It will."

"How can you be so sure?"

"Because it always does," I said. There was another reason that I was anxious to explain. You know how flight instructors are. I pointed to the eastern sky where there was a bank of cirrus clouds. The sun was within five to ten minutes of rising above those clouds. "See how the sun is about to pop out from behind those clouds?"

"Yes," Paul answered.

"When it does," I told him, "the sun will burn off this fog in short order."

"Hmmm," Paul allowed.

"There's one other reason I'm not worried," I said, pointing out my window to the west. Way over, almost at the range of where we could see, the ground was dark instead of white. "If we had to, we could go over there where it's not foggy and land."

"But what if the engine quits?" Paul wanted to know.

"Then I guess we'd do the same thing we'd do at night. Ride it down, cushion the landing, and hope for the best." It seemed reasonable to me.

Now, here's where it's fun being the IP. The sun rose above the cirrus layer and burned off the clouds. Imagine that. Lucky guess on my part!

Busting a Check Ride

Date: 12/20/74
A/C Type: UH-1H
Registration #: 70-16247
Route of Flight: Fort Bragg – Duke – Fort Bragg

Don Calderwood came on board the 57th as our Instrument Examiner. It was a slot that had not been filled before, but now we had someone in the unit authorized to give us the instrument check rides that were required for us to remain current.

It was the time of year when we flew a lot of IFR, so Don decided to give me my instrument check ride on an actual mission. He rode in the jump seat while we made an IFR run to drop a patient off at Raleigh-Durham and return. It was about 10:00 at night when we arrived back at Simmons and asked for a GCA (Ground Controlled Approach). A GCA is a military thing, though civilians can ask for an airport surveillance approach at many facilities, and there are similarities. On a GCA approach, the controller provides both vertical and lateral guidance to the runway using radar. He gives verbal commands to the pilot to "turn left, turn right, begin descent, above glide path, increase your descent, stop descent," and so on, guiding you right to the ground.

Instrument approaches have minimums. They have minimum weather conditions below which you're not supposed to commence an approach, and they have minimum altitudes below which you're not supposed to descend if you don't see the runway. Busting minimums is not only illegal, it can and often does get a body killed. GCA approaches typically have lower minimums than other kinds of approaches. That was good because Fayetteville Approach Control had just told us the weather at Simmons Army Airfield and it was not good. Not below minimums yet, but getting there fast.

I began the approach with every bit of confidence that we would soon be on the ground. The controller gave me the verbal commands every few seconds and quite often they were "on course and on glidepath." I knew I was doing a good job on the check ride. Then the controller said, "Over the runway threshold, take over visually and land." I didn't see any runway threshold. I asked the copilot and he didn't see it either. Neither did Don. I executed a missed approach, which is the procedure you do when you don't see the runway at minimums. A missed approach involves climbing back to a safe altitude and sorting out your options about what to do next. After asking about the weather at nearby Pope Air Force Base and Fayetteville Airport and finding out they were reporting weather no better than Simmons, I made the command decision to try another approach.

The second time around was the same as the first, a beautiful approach, on course and on glidepath most of the way, but at the decision height, no runway. During the missed approach, I asked Fayetteville Approach again for the the weather at the Fayetteville airport. Sky obscured, visibility a quarter of a mile, he told me. Below minimums.

"We're getting really low on fuel, you guys," I said to Don and the crew. "I'm not inclined to go to Fayetteville when this controller can talk us all the way to the ground." He could, too.

So, we went around and did it again, and sure enough at minimums I didn't see anything. I started executing a missed approach, which involved a climbing left turn. As I began the turn I caught sight of a light on the ground. It was one of the airfield perimeter lights. "Runway environment!" I announced and turned back to the right where the runway had to be.

We found it and followed the white center line to a taxiway, then hovered to parking and shut down.

"Well, Freeman," Calderwood told me, when we got out of the helicopter. "You busted your checkride for busting minimums, but thanks for getting us on the ground in one piece."

"Busted my check ride? Busted minimums? I saw that light."

"You can't show me anywhere in either the FAA regs or the Army regs where an airport perimeter light can be construed as runway environment." Calderwood justified.

"Come on, Don, you don't want to go through this again," I pleaded.

"Can't do it," Don said. "You can't go around busting minimums."

"What would you have had me do?" And so the argument continued. I don't remember if he really failed me or not, but it was no big deal. I'd simply have to go fly with him again.

To Every Thing There is a Season

Toward the end of 1974 I was becoming disillusioned with how things were going in the Army. I had applied a couple of times for the bootstrap program that would have allowed me to finish college, but each time I was turned down on a technicality. For two years the MAST mission had been very rewarding, a source of tremendous job satisfaction. But now it seemed the brass in the Army had rather play war games than save lives and MAST missions were becoming a low priority. This caused the civilian medical and law enforcement communities to quit calling upon us because we could no longer be counted on to be there.

I loved being the 57th's Instructor Pilot. The pilots in our unit were considered the elite among the Army pilots at Fort Bragg and here I was with the rare privilege of being the professional among professionals.

My logbooks for the time I had been flying at Fort Bragg revealed an impressive number of lifesaving missions I had been privileged to fly. Among them were:

- 19 patients with head injuries or brain tumors

- 29 premature babies

- 10 patients with internal injuries from automobile accidents

- 59 patients injured from parachute jumps

- 17 patients with severe neck or back injuries

- 2 victims of poisoning or drug overdose

- 10 victims of gunshot wounds (most of them self-inflicted)

- 3 victims of child abuse (one of them died at the hospital)

- 5 motorcycle accident victims paralyzed with neck or head injuries

- 2 patients with acute respiratory distress (one elderly, one young)

- 4 severe burn victims, two of them from an Eastern Airlines crash at Charlotte
- 1 severe electrical shock (he died)
- 2 known suicide attempts (slashed wrist, overdose)
- 1 traumatic hand amputation (hand was successfully reattached)
- 7 cardiac arrests
- 25 more with everything from cancer, kidney failure, leukemia, fractured femurs, etc.

It really irked me that the Army was playing games. At least I got to give 90-day standardization check rides to the other pilots. But I was going through a stage where I had to do something productive each night after I got home in order to feel like I had accomplished anything that day.

The kicker was that I was coming up for an unaccompanied tour to Korea and Joyce was expecting our first child. We talked it over and decided that getting out of the Army was a good idea. I started the paperwork in January, 1975.

Nap-of-the-Earth Training

Vietnam was a thing of the past and the people in charge of Army doctrine had decided that the next war would be in Europe or the Middle East where helicopter survivability would depend upon being able to blend in with the natural terrain and vegetation. Nap-of-the-Earth flight was invented.

In reality, pilots have been flying nap-of-the-earth (NOE) since the Wright brothers. We just didn't call it that. We called it "low-level" flying. Now that it was a doctrine and had a fancy name, it was legal and we were all going to have to learn how to do it officially.

The IPs were the first to be trained. Then we were to train the other pilots in our respective units. It sounded to me like as good a way as any to spend my last six or seven months in the Army.

Some IPs from Fort Rucker came up and put all of the Fort Bragg IPs through an NOE qualification course. The syllabus included hovering out of ground effect and hovering downwind; a manuever called the NOE quick stop, which was basically decelerating at low level without putting the helicopter's tail rotor in the grass or trees; plus masking and unmasking, which was hiding behind something, popping out to shoot, then hiding again. Then there was low-level navigation, which proved to be the only challenging part of the course.

Think of perspective. In flying, altitude gives you a better perspective on things from a navigation standpoint. On a short trip, given enough altitude, you can see both the beginning and end of the trip from the same location. Even on a long trip altitude allows you to see landmarks far off into the distance, so you don't have to concentrate on navigating each little part of the journey.

Go down on the deck, however, and navigation becomes something entirely different. Then factor in variations in speed. In this new Nap-of-the-Earth, hide behind the terrain and trees kind of flying there were times when you crept along slowly and other times when you ran flat out. We were all used to flying with a 90 knot finger moving across the map. Now, it might have to move at 40 knots, then 110, then 60, then 20, then back to 40. We were forever getting ahead of or behind ourselves on the map until we figured out how to really read the terrain.

In normal navigation you use radio navigation aids, supplemented with manmade features such as roads or railroads, water towers, etc. Now we had to ignore such things since they might be destroyed or altered in combat. We had to learn how to read topographical maps so that we could recognize valleys, draws, saddles, ridges, peaks and other such terrain features as we flew along. I was sure some infantry commander thought of this entire doctrine and was laughing himself silly watching a bunch of aviators learn how to read a topographical map.

We stayed lost most of the time during the first week of our training. Thankfully, by the time I started teaching others to do it I was getting the hang of it myself. By the second week of instructing it was becoming second nature.

From February until the middle of June I averaged four hours a day instructing other pilots in NOE techniques. It is amazing how we adapt to new things. I got to a point where I felt like I could fly a Huey through any opening it would physically fit through at any speed up to VNE (velocity not to exceed, i.e., top speed for the aircraft, which for a Huey is 120 knots). While the newbies were struggling to follow a course at 30 to 40 knots (just as I had struggled a few weeks earlier), now I could fly it at 110 knots and never miss a turn. It was a shame, I thought, that all of that talent was going to be wasted. Just about the time I got really good at it, it was time to leave the Army. I remarked to Jim Miles one day that I felt like a Huey was an extension of my arms and legs, that I felt more comfortable flying a Huey than I did driving a car.

"That's only natural," he told me. "Think about it. You've spent an average of four or five hours a day in the pilot's seat of a Huey for the past five years. That's a good deal more time than you've spent driving a car, I'd guess."

Good point. I hadn't thought of it that way.

Civilian Flight Instructor

When I knew I was serious about getting out of the Army, I began to think about what I would do for a living. There was no doubt in my mind that I wanted to continue flying and one way to do that was to become a Certified Flight Instructor, a CFI. I was a member of the Fort Bragg Flying Club and still had some GI Bill money left, so I started working to become a CFI.

J. D. Huss was a pilot's pilot. He had been an Army Aviator himself until a year or two earlier. Now he instructed full time at the flying club. When he was in the Army, J. D. had flown everything from light observation helicopters to twin-engine turboprop Mohawk surveillance airplanes. He was building time for an airline pilot job, but that didn't keep him from being an excellent instructor. It was easy to tell that J. D. loved instructing and he was very good at it. When he took on the task of teaching me to be a flight instructor, J. D. had a little over 5,000 hours himself. He had just bought an airplane, a 1939 Monocoupe. His life was totally wrapped up in flying, so I was being tutored by a master.

My first CFI lesson with J. D. was on April 23, 1975. We went up in a Cessna 150 with me in the right seat, which is the seat a CFI normally occupies and J. D. flying the left seat. He began teaching me how to demonstrate and explain the basic flight maneuvers—turns, climbs, glides and straight and level flight.

I discovered right off that you really have to know a subject well in order to teach it, especially in such a hostile environment as the hot, noisy, and cramped cabin of a small training airplane. J. D. knew his stuff and over a period of several months he made sure that I knew mine. We held detailed discussions before and after every flight. During those discussions we dissected each maneuver required for the private and commercial check rides. Using a model airplane and a blackboard, I learned how to describe how each maneuver was performed and how the basic forces of thrust, drag, lift and gravity came into place during the maneuver. I had to be able to describe this to a student and answer any questions he or she might have about what I was describing.

We discussed torque, P-factor, control surfaces, airspeeds, and power settings. Each piece of the training puzzle had to be thoroughly understood, including an understanding of how students learn, how to keep them motivated, how to gauge their understanding, on and on.

When it came to flying the maneuvers, J. D. made sure that I could not only fly the maneuver precisely, but could explain it while doing so, and from the right seat, which was a little awkward at first. J. D. was patient and thorough. When we worked on short and soft field takeoffs and landings, we didn't just work within the guidelines of the maneuver as described by the FAA. We worked toward extracting maximum performance from the airplane based on the performances charts and procedures defined in the airplane's pilot operating handbook (POH).

When we worked on the commercial maneuvers—the chandelles, lazy eights, and pylon eights, J. D. wasn't satisfied and taught me not to be satisfied until they were flown as smoothly and precisely as possible. We flew a couple of afternoons a week, and the rest of the time I was doing NOE training and an occasional Dustoff or MAST mission. I was determined to log as much time as I could during my last few months in the Army.

In early June I took and passed the Flight Instructor written exam and J. D. recommended me for the flight test. I flew to Raleigh on the appointed day, but the weather didn't cooperate when it came time to fly the check ride. I passed the extensive oral portion of the test, including preparing a lesson plan, which I was supposed to fly with the examiner during the check ride. The weather never cleared and we had to postpone the flight exam until a later time. That later time didn't happen until after Joyce and I had moved to Mississippi.

Chapter 4 – Hometown Pilot

After leaving the Army, Joyce and I, along with our young son Jamie, moved to my hometown, Oxford, Mississippi. I joined the National Guard as a helicopter pilot, finished college using the GI Bill and worked at the local airport as a flight instructor and charter pilot. During this two-year period, we became part owners in a couple of different aircraft.

As soon as I knew for sure that I was getting out of the Army, I had contacted the Mississippi National Guard to see if they had any openings for pilots. I was assured they did. What I really wanted was a full-time job with the National Guard. Those were civil service jobs with good pay and good benefits and I figured that with my qualifications as both a Maintenance Officer and an Instructor Pilot that I would be a shoe-in for such a job. What I didn't figure on was the politics involved. Mississippi had a well-entrenched "good ole boy" network and I just didn't have the political connections to get such a job. I was assigned to a unit, however, and began flying weekend Guard drills right away.

I started working as a CFI, a Certified Flight Instructor, at the University-Oxford Airport. In addition to flight instructor duties, the job included being an on-demand charter pilot. In the National Guard I became qualified in the OH-6 observation helicopter as well as the Huey. Then I was put on IP status with the Huey and shortly thereafter became an OH-6 IP as well. The minimum annual flying hours for a National Guard pilot is 60, meaning a pilot is required to fly at least 60 hours a year to remain qualified as a pilot. If you're an instructor pilot, the minimum is doubled to 120 hours. Since I was an IP in two different aircraft, my minimum number of annual hours was 240. While this isn't much for an active duty pilot, it is a considerable amount of flying for a "weekender."

In order to get my hours in, I flew a lot. Basically, I flew nearly every weekend plus many Wednesday nights. It got to a point where I was allowed to pick up an OH-6 at the Armory in Tupelo on Friday night and keep it all weekend. Often I would get an assignment to fly one of the Generals here or there, or to participate in some training exercise, but just as often I was on my own. Get your flying hours in I was told. What more could a fellow ask for, a government helicopter and a government credit card.

The extra flying in the Guard came in handy for another reason—money. It was taking three sources of income—the GI Bill, National Guard pay, and the income from instructing and flying charter flights—to make ends meet. Joyce was just as much a "pilot's widow" as ever, since I was again on call pretty much 24 hours a day, seven days a week, just like when flying Dustoff or MAST missions.

The flying school at Oxford had active ROTC programs going on for the Air Force, Army, and Navy. My military training experience fit right in, since the training syllabus for these ROTC cadets was military oriented. I managed to fly three or four lessons a day and found that I really enjoyed the challenge of teaching primary flight students to fly.

I had other students as well. Some were local college students from well-to-do families, and some were local businessmen or women who just wanted to learn to fly.

The charter flights were all flown in single-engine airplanes, since the air service didn't have a twin. Usually it was a Cessna 172, but on some occasions the flight was made in a Cessna 210. I flew lawyers to depositions, college professors to speaking engagements and made parts runs for local manufacturing companies. I also flew for the US Forestry service on a variety of missions.

The Luscombe

Herschel Lamb, a local mechanic, had restored a pretty little airplane. It was a 1947 Luscombe 8E, all red, white and blue. But Herschel was about half scared of the airplane. Truth be known, I think he was fully afraid of the airplane. Luscombes have a reputation for being squirrelly on the ground because they're a tailwheel airplane with a narrow landing gear. It's a groundloop waiting to happen; that's what the oldtimers said about her. I wanted to fly the Luscombe so badly I could taste it.

I got Herschel to take me up in the Luscombe, but he wouldn't let me take off or land. Sitting there in the right seat, I couldn't help but notice that he was making a bunch of mistakes in his handling of the plane. The instructor in me was itching to take over, but Herschel hadn't hired me to give him any dual. He was just being gracious and taking me for a ride.

He made his approaches way too fast, so naturally the little plane floated halfway down the runway before touching down. When I suggested to Herschel that he try his approaches a little slower, he told me he didn't want to stall the plane. We were approaching at 70, sometimes 80 knots. I was pretty sure the stall speed was down around 45 or 50. I began to think that I'd have to buy the airplane to get to fly it.

Somebody else bought it first, and the new owner hired me to teach him to fly it. The first order of business, I told the new owner, was for me to learn the airplane myself. With his blessing I began to do so.

Flying it was a joy. I had no trouble with the takeoffs or landings. I'm sure being a helicopter pilot helped. There was no manual for the airplane, so I went to the library to learn about the plane. I dug up pilot reports and advertisements from 1947 magazines and learned what the recommended airspeeds were for various phases of flight and what the aircraft's limitations were. When I learned that the power off stalling speed was 42 knots, I began flying my approaches at 60 knots, then at 55. I found the Luscombe to be quite a short field aircraft at those approach speeds.

I began giving instruction in the Luscombe and soon its new owner decided he wanted a bigger plane. I got some friends of mine to enter into a partnership with me and we bought the Luscombe. I taught myself to do loops and snap rolls in it and began teaching my partners and their wives to fly it. I also used it as transportation to Guard drill on the weekends.

Instructing in the Luscombe was a bit of a challenge. It only had brake pedals on the left side of the cockpit and those were hard to use. They were heel brakes, instead of the toe brakes found on modern airplanes. The fact that it was a taildragger made it difficult to taxi, especially if there was any wind. Since the landing gear was designed such that the main wheels were close together, the airplane was susceptible to ground loops and had to be watched closely all the time.

I figured the flight instructors in the forties and fifties taught people to fly in Luscombes, so I could, too. I taught a number of people to fly in that airplane and at least one of them took his check ride for his license in the Luscombe. Several of the others soloed in it. The time spent in that right seat definitely helped to sharpen my reflexes and judgement.

The Humbler

Date: 9/19/75
A/C Type: Cessna 150
Registration #: N18152
Route of Flight: University-Oxford (UOX) – Local

I was pretty good at recognizing a student pilot with a chip on his shoulder, since I had carried one around myself during my early flying career in the Army. One such chip belonged to an ROTC cadet who was part of a rather famous flying family. His father and older brother were both airshow performers. The dad flew an F8F Bearcat and the brother dazzled crowds with his tiny Acrojet.

My student had never flown as a pilot before. Yet he seemed to think that because his father and brother were well-known and accomplished pilots, that he already knew how to fly without having to learn. He apparently thought skills had been imparted to him by osmosis. I couldn't teach him anything. But his flying was atrocious. I decided to get his attention with something I called the "humbler."

During one of our flying lessons I had the young ace take us up to a safe altitude where we could do a stall series. I had him slow the Cessna 150 down and put down 20 degrees of flaps. Then I talked him into a medium-banked turn with the aircraft out of trim, i.e., the ball not centered in the turn and bank indicator. I had him keep adding back pressure to the control yoke. I kept him pulling until the 150 broke into a stall.

From in the cockpit it seemed the plane had flipped over onto its back and was about to enter a spin. That's what a Cessna 150 will do in an accelerated, out-of-trim stall, so it was no surprise to me. It sure caught the ace by surprise though. He started fighting the plane and trying to straighten out the turn with ailerons. That's the wrong way to recover from an accelerated stall. Things only get worse, which is what happened. The little plane was eating his lunch. Finally, after what must have seemed to him like an eternity, I said to him, "Turn it loose." He looked over at me with panic in his eyes, still fighting it. "Turn it loose!" I said again, emphatically this time. He released the controls and the airplane stopped its twisting and turning, recovered from the stall and flew along in level flight, as gentle as a kitten.

"This airplane can fly itself better than you can," I told him and apparently he believed me. Our lessons after that little episode were productive and he wound up making a pretty good pilot.

Jane's Crosswind Landings

Date: 9/29/75
A/C Type: Cessna 172
Registration #: N1388U
Route of Flight: University-Oxford – Local

It had crossed my mind that Jane would never solo. She was patient enough, but there were certain concepts she just couldn't seem to grasp. One was the effect of crosswinds.

On this particular day, I was frustrated and about to call an end to our lesson. Then I got an idea.

"Jane," I told her, "you know how I normally don't let you use more than the first 1500 feet of runway?"

She looked at me quizzically.

"Today we're going to try something different," I said. "This time you have the whole 4,000 feet. But I don't want you to land. Instead, fly a normal traffic pattern—like if we were doing touch and goes—but instead of landing, I want you to fly down the whole length of the runway at about ten feet of altitude."

Jane didn't say anything. She just looked at me like I was crazy.

"Go ahead," I encouraged her. "It will be a good learning experience."

I had used the right word. To Jane, the whole flying thing was about experience. Since her husband had died and left her with a healthy bundle from his insurance proceeds, Jane was out to experience all life had to offer. It was all Jeff Meaders and I could do to stop her from buying a new Mooney before she found out whether or not she could learn to fly it. We were in her twenty-sixth hour of dual instruction and I had not yet felt comfortable letting her solo.

She could fly a decent pattern, she just couldn't land—at least not consistently. It was worse when there was a crosswind, as there was on this day. The wind was blowing across the runway at eight to ten knots. That was well within the capabilities of the airplane, but it was giving her fits.

She was okay with the crab angle down final. It was when she attempted to line up for touchdown that she lost it—every time. Or, at least nine times out of ten.

"That's right," I coached her, as we neared the ground and she started her flare. "Leave some power on. Not a lot, just enough to keep us flying. Now just fly right down the center of the runway, all the way to the other end."

"Crab or slip?" she asked.

"Crab, this time," I answered, impressed by the fact she knew enough to ask the question. Jane understood the techniques. She just couldn't seem to do them.

I held my breath as she leveled out just over the runway. I had to help her get the power setting right, but then we were flying down the runway and SHE DID IT! She flew all the way down the runway, staying more or less over the centerline. Sure, she had to make some heading adjustments, but she saw what was happening and responded to it.

"Let's go do it again," I told her as I added power for us to climb back up to pattern altitude.

As we turned final the second time around, I instructed her, "This time use a slip and keep the nose aligned with the runway."

The slip was a little tougher and she struggled at first, but by the time we were halfway down the runway, she had figured out the right amount of wing to hold into the wind and she was keeping the nose aligned with the direction of flight.

"Again," I said as we climbed away from the runway. "Another slip."

We flew the traffic pattern and then flew a slip the full length of the runway again. This time it was hard for Jane to contain her excitement. She was grinning from ear to ear, because she knew she was getting it.

The fourth time as we were flying level over the runway, I eased the power off and the plane settled to the runway. Jane seemed surprised when the wheels touched, but she kept her cool and continued to hold the nose off and use the controls properly to compensate for the crosswind. We rolled a little way and I gave the command, "Let's go." Jane fed in the power, and automatically checked to make sure the flaps were up—a routine touch and go.

She waited until we were well established in the climb before she turned to me with a big grin on her face. "I just did an almost-perfect crosswind landing, didn't I?"

"You sure did," I affirmed.

"Why didn't we do this before?" It was a good question.

"Because I just thought of it," I replied.

It would have helped me if one of my flight instructors had thought of an exercise like that when I was learning to handle crosswind landings. By making Jane fly over a relatively long runway while countering the effects of a crosswind, I stumbled onto a method of training that would allow her plenty of time to analyze what she needed to do. A new training tool had just been added to my repertoire.

Three Military Pilots in the Soup

Former Mississippi Lieutenant Governor Charles Sullivan flew a Cessna 310 that belonged to North Mississippi Savings and Loan Association. He was a lawyer and a businessman and Chairman of the Board of the S & L. In his other life Governor Sullivan was an Air Force Reserve General.

There came a time when the governor wouldn't be able to make some upcoming trips, so he enlisted the aid of Colonel Brandon to fly them for him. Colonel Brandon was ex-Army and had been Ole Miss's pilot for a while in their Cessna 402. He was more than qualified to fly the 310. I wanted to fly the trips, but didn't have enough twin-engine experience for the job. The general and the colonel were very gracious to me and inivited me to fly along in the back seat the day the general checked the colonel out in his airplane.

So there we were, three ex-military aviators with a considerable amount of flying experience between us, going for a ride in the Cessna 310 on a cloudy, low-ceiling day. The weather wasn't much of a concern since both pilots flying were highly experienced weather veterans. During my medevac days I had racked up a pretty impressive number of instrument hours myself, so riding in the back seat while in the clouds didn't bother me a bit.

They filed a round-robin IFR flight plan from Oxford to Clarksdale and back. We took off at Oxford and went into the clouds at about 800 feet above the ground. The Colonel was flying and the General was in the right seat. We leveled off at 4,000 feet and the Colonel turned on the autopilot. He started asking questions about various items on the panel and the radios and such and the General explained the peculiarities of his particular airplane, one of which was that it wasn't going as fast as it should be going.

"Hmm," the General wondered, "I wonder why we're only doing 140 knots? We should be going faster than this."

"How about the power settings?" the Colonel asked, and the General answered that no, they were about where they should be.

It took around thirty minutes to get to Clarksdale and from time to time along the way, one or the other of them would comment on the airspeed. I looked on from the back seat and doggone if I could see any reason for it.

Then came time for the prelanding GUMP checklist (Gas on the right tanks, Undercarriage [landing gear] down, Mixtures rich and Props set). When the Colonel reached to put the landing gear down was when we all discovered that the gear was already down. It had never been put up. There was our explanation for the slow airspeed.

Sixville

One of my all-time favorite airplanes is the Cherokee Six. My love affair with the Six started with a 1966 model—N3547W. Four-seven-whiskey belonged to Jim Cregar, a local businessman who knew the value of owning an airplane, but wasn't particularly enamored with flying it. The story I heard was that while crossing Lake Pontchartrain on a hazy day, Jim lost the visible horizon and got a bad case of vertigo. If he hadn't had an autopilot, he'd have been in the drink. That's just what I was told. The story suited me because it meant of lot of opportunities for me to fly the Six.

Jim needed someone to fly occasional trips for him and I needed access to an airplane from time to time. The arrangement that suited us both was that I became his pilot and in return he let me use his airplane as if it were my own. It was an arrangement I really liked, so I was careful not to abuse the privilege.

Four-seven whiskey wasn't a pretty airplane. It was hard to tell what color she had been when she left the factory, but my best guess is orange and white. The darker color was kind of a faded pink by the time I met her and the white was dull and rubbed off on your hands.

There were many things I liked about four-seven-whiskey and a few I didn't. On the plus side you could fill the fuel tanks, fill all six seats with people, even grown-up people, and go somewhere—legally. Takeoff performance was good even with the seats full and the cruise was an honest 140 knots. The throttle was a big old knob that you could grab a fistful of and when you pushed it forward be rewarded with the powerful sound and feel of the big six-cylinder 260 horsepower Lycoming engine under the hood. The starter was a pushbutton, just like on the old Ford trucks. I don't know why I liked that, but I always have.

On the negative side, the directional gyro was one of those horizontal jobs that turns the opposite direction from what you would expect it to. Plus, it precessed at something like three degrees a minute.

Still she was a joy to fly and I have a few memorable trips in her to tell you about. Later I rented a Cherokee Six on occasions, then flew one for a company and eventually owned a fourth interest in a 1976 model for a short while.

Livestock Navigation

Date: 10/3/75
AC Type: Cherokee Six
Registration #: 3547W
Route of Flight: Oxford – Little Rock – Winona – Oxford

Jim Cregar needed me to take a package to Little Rock in his Cherokee Six. I took along one of my commercial students and we headed off late in the afternoon. The flight to Little Rock was routine. The trip home, however, was after dark and contained a few surprises.

We departed Little Rock VFR and leveled off at 3,000 feet, heading generally east. When I say generally, you have to remember that this airplane had an old directional gyro that was totally unreliable. Finding Little Rock had been easy enough because there is a VOR real close to the airport and it was daylight. I figured finding Oxford would also be pretty simple, even if it didn't have a VOR. My plan was to head east and keep Memphis on our left and Greenville on our right. That would keep us going pretty much in the right direction. When we crossed the Mississippi River, we'd start looking for I-55 and when we crossed it, we'd look for the town about twenty miles ahead with an airport just northwest of the city a mile or two. I knew where the airports were for all of the towns anywhere close around and none of them were northwest of the city, except for Oxford.

The first little challenge came after we leveled off and I switched on the autopilot. I decided to burn fuel out of the left tip tank. The fuel tank selector was at the base of the center pedestal, which put it on the floor between the two front seats and it wasn't lighted. I reached down and moved it to the left. I was rewarded for that simple gesture with two sputters from the engine, then silence.

Have I mentioned before that it was a dark night? There was no moon and there were clouds above us blocking even what light the stars might have provided. Fortunately my engine-out emergency training kicked in and as I went through the engine failure drill, moving the fuel tank selector back to the right made the engine start running again.

There should have been some kind of detent or something to keep you from inadvertently moving the selector to the off position, when all you wanted to do was draw fuel from another tank, but there wasn't. I shoved it over to the right some more, just so it wouldn't be anywhere close to the off position and the engine kept running.

The Mississippi River showed up about when I expected it to, and the lights of Memphis were still on the left and the lights of Greenville were apparently somewhere down there to the right. I tried to identify the towns that were close around us, but didn't have much luck. I wasn't worried though because I-55 was still ahead and I was confident that I would know the towns along it when we got there.

It didn't quite work out that way, however. We crossed the interstate and the town that should have been Batesville obviously wasn't because the airport was in the wrong place. The town up ahead wasn't Oxford, either, because its airport was also in the wrong place. I got a bearing off the Holly Springs VOR, but it didn't help much because I couldn't get another VOR bearing to cross it with.

I figured we'd just turn and fly along I-55 until I came to a town that I recognized, but I wasn't sure which way to turn. I chose south. When we came to a town, I still didn't recognize it, but it had an airport and I knew we'd be safe landing there without talking to anybody, because none of the small towns in Mississippi had a control tower.

On landing, the windows all fogged up, making for an interesting rollout. I opened the door a crack so I could see outside and my student started wiping the windshield with his sleeve. Only it was fogged up on the outside, too, so that didn't do much good. I managed to find the taxiway and the ramp and as soon as I shut the engine down, I knew exactly where we were. The mooing cattle sounds were a dead give-away—Winona. Winona had a cattle auction right beside the airport. No other airport I knew of had that.

I didn't even bother to check the water tower after hearing the cattle. We got the defroster blowing to clear the windshield and took off. I knew how to find Oxford from Winona, directional gyro working or not.

Professors to Meridian

Date: 10/25/75
AC Type: Cherokee Six
Registration #: N3547W
Route of Flight: Oxford – Meridian – Newton – Oxford

It was on a cold, windy October day that I had the opportunity to fly a group of college professors to a seminar being held at a junior college in Newton, Mississippi, near Meridian. We left early in the morning in 47W, Jim Cregar's Cherokee Six. The ceiling was low overcast, so I had filed an IFR flight plan. We went into the clouds right after takeoff.

I did a climbing turn over the airport to get enough altitude to pick up the Holly Springs VOR and intersect a radial that would take us in the right direction. I was proceeding on course just about the time Memphis Center established radar contact with us.

Within just a few minutes I began having trouble with the directional gyro. It was drifting about 15 degrees every five minutes, which made it very difficult to maintain a correct heading. Fortunately I was able to maintain course by staying on my assigned VOR radial. I was sure hoping I wouldn't have to fly an instrument approach at Meridian.

The passengers were oblivious to my plight and were very trusting, even though when they looked out the windows they saw nothing but thick, gray clouds. By the time I got to Meridian it was VFR, but just barely. We landed safely and since the weather was forecast to clear by afternoon, I told the guys I would pick them up at Newton when it was time to go home.

I landed at Newton about 4:00 in the afternoon under a 1500 foot ceiling. So much for the weather clearing up. There was a highway that led from Newton to Oxford and there was relatively good visibility under the clouds, so I decided to follow the highway home, rather than try to fly IFR again with the bad gyro. Oxford was only an hour and fifteen minutes away.

The first part of the flight went well, though a little bumpy, but by the time we were forty miles out from Oxford the ceiling had begun dropping. The bottom of it was really ragged and in places the visibility was obstructed by even lower clouds—scud. The ceiling finally got so low that I had all the passengers watching for radio towers while I concentrated on staying over the road. I seriously began thinking about climbing up through the clouds and contacting Memphis Center for a clearance into Oxford. The fact that I didn't trust the gauges was my only deterrent.

It turned out to be a good thing that I didn't pop up and try to get an IFR clearance. Ole Miss had played LSU that day and a couple of hundred planes had flown in for the occasion. Everything from Lear Jets to King Airs to piston twins were lined up on the sides of the runway, engines running, waiting for IFR releases out of there. There was a mobile FAA control tower in operation that day because of the traffic.

Within twenty miles of the airport we were beneath a 700 foot ceiling, and visibility had improved so that I could see at least 10 miles or more. I was in uncontrolled airspace with more than a mile visibility and clear of clouds, so technically, I was legal, but I wasn't sure how that fit into the picture with a temporary air traffic control area in affect at the Oxford airport.

Monitoring the tower frequency I heard a steady stream of airplanes calling for IFR clearances. I wondered how I would possibly get in VFR if the airport was IFR and so many planes were waiting to get out. The tower explained to the departing traffic that the reason they were having to wait was that the airspace above them was already saturated with IFR traffic.

Nervous, I keyed the microphone and called the tower fifteen miles out for landing clearance.

"Three-five-four-seven-whiskey, Oxford Tower, are you VFR?"

I wasn't a hundred percent sure my answer wouldn't get me in trouble. I was VFR, but was I legally VFR? I took my chances. "Four-seven-whiskey, yes, I am," I answered.

"Four-seven-whiskey, cleared to land, runway nine."

Was that all there was to it? I was still fifteen miles from the runway, flying northbound and fighting a very strong north wind and I was cleared to land with all of that heavy iron waiting to get out. Those pilots must have resented me something fierce.

It took forever to get to the airport. When I finally lined up for final, I could see at least twenty jets and turboprops with engines running, waiting to depart. All of that jet fuel and here I came in this little Piper, bopping along under the clouds. I was sure they all thought I was the reason they were being held on the ground. But that wasn't the case at all. Because there was a tower there that day, there was an Airport Traffic Area, which made it a requirement that I get a clearance from the tower prior to landing. But, there was not a Control Zone at the airport, so the three miles visibility and 1,000 foot ceiling limits did not apply. I did not have to obtain an IFR clearance to land at the airport, even though the weather was below VFR minimums for controlled airspace. The airspace below 1200 feet there was uncontrolled. It's amazing how many rules there are for pilots to know and understand. At times it gets confusing.

Fire Patrol

Small town pilots take on any kind of flying job they can find. You have to if you want to make a living doing the thing you love. Jeff Meaders, the airport manager and Fixed Base Operator at the University-Oxford Airport had been making a living in the flying business for many years. He knew how to combine fuel sales, aircraft rental, flight instruction and charter operations into a successful business, even in a small town like Oxford. One of the things he did to keep us flying was to contract with the US Forestry Service to provide fire patrol and other services.

Flying for the Forestry Service required that you take an annual qualification check ride with one of their check pilots. This was in addition to maintaining your status as a charter pilot through periodic FAA check rides. The Forestry Service pilot that flew in one day to check me out as a Forestry Service pilot was a guy I knew from my Army days, Charlie Morris. In fact, in my closet at home was an army flight helmet that belonged to Charlie. He had left it in my car a couple of years earlier in North Carolina and for some reason I had never gotten it back to him. I went home to get it and Charlie sure was glad to get it back. We flew around a little in a Cessna 172 doing some slow flight and stalls and Charlie decided that I was okay to fly for the government again.

Fire patrol is what I did most often for the US Forestry Service and I'll tell you it was not the most exciting flying I ever did. A forest ranger would fly along and we would fly in slow circles over the national forest for the better part of an afternoon looking for fires. Whenever smoke was spotted, the ranger would call someone on his handheld radio and they would dispatch a team to take care of it.

It was so boring that I started trying to spot fires before the ranger saw them. Nine times out of ten, when I did point one out to him, he'd say, "oh, that's a prescribed burn," and pay no attention to it at all. A prescribed burn is when the forestry service decides that the wind, temperature and humidity conditions are such that if they set a fire, it will burn out the underbrush, but not get hot enough to catch the big trees on fire. Burning out the underbrush is apparently good for the forest, so they do it from time to time.

I was pretty good at reading the 1:50,000 topographical map the ranger used to pinpoint the fires. There were times when he would be twisting and turning the map, trying to orient it to where we were so he could locate where the fire was, and I would reach over, point at the map and say, "here, we're right here, and the fire is right there." He asked me how I could do that so easily, and I explained that while flying Nap-of-the-Earth in the Army I had learned to read one of those maps while flying at various speeds right over the treetops. Once you learn how to do that from just over the treetops, looking at it from 1500 feet up is nothing. Perspective is an amazing thing.

We didn't just spot fires for the Forestry Service, and we didn't just work the Holly Springs National Forest, which was near Oxford. Sometimes we flew down to Rolling Fork or Yazoo City in the Mississippi Delta and flew over the Delta National Forest to assess the damage caused by beavers building dams and flooding certain sections of the forest to the point that it killed off the trees.

Most of the time when I flew for the US Forestry Service it was with a ranger aboard and the mission was to spot forest fires and direct the ground crews to them. During what they called "fire season," mid-to-late summer when there hadn't been a lot of rain we often stayed over the National Forest most of the day. One time we used the airplane to catch an arsonist.

During the summer months of 1976, someone had obviously been setting fires in the Holly Springs National Forest. There was too much of a pattern for the fires to have been coincidental. The Forest Service had a suspect, an individual who had expressed his contempt for the government because of a land deal that hadn't gone his way, but there was no proof.

The Chief Ranger and I circled above the forest all afternoon one day when we suspected he might strike. With the engine throttled back and the aircraft droning slowly through the sky, we were burning such a small amount of fuel that the Cessna 172's endurance was longer than ours.

We observed the suspect leave his house and drive into the forest in his pickup. Rangers were located at strategic places on the ground and we alerted them to be ready to set up roadblocks. Sure enough, one fire after another began to spring up along the road the man was driving. It was obvious from the air that he was leaving a trail of small blazes behind him. We radioed the rangers below and they closed in. The man drove around a curve straight into a roadblock. Before he could turn around, rangers rushed his pickup and caught him with several Molotov cocktails on the seat beside him. He had been driving down the road throwing them out. I learned later that our evidence convicted him.

We also observed an ongoing affair between a farmer and what appeared to be another farmer's wife. Every afternoon she would leave her house at a certain time and he would leave his house and they would meet on an isolated road, where she would get out of her car and into his pickup. They'd be there for an hour or two, then separate and go their own way. It was just coincidence that we saw them the first time, but then we started watching them and they were a regular thing. It wasn't any of our business, but watching their rendevous helped break up the monotony of the surveillance flights. I'm sure they never knew that big brother was watching them from the sky.

Demonstrated Crosswind Component

Date: 3/4/76
A/C Type: Cessna 210
Registration #: N1EG
Route of Flight: Atlanta – Oxford

A cold front had passed through earlier in the afternoon leaving clear skies behind. When Randy and Ron finished their business in Atlanta, I felt comfortable heading for home. It was late and there was a strong head wind between Atlanta and Oxford, but otherwise I expected the conditions to be fine. We departed Fulton County Airport with my wife, Joyce, in the right front seat holding our infant son, Jamie, and our two friends, Ron and Randy, in the back.

What a view during the climb out! Since the rain showers associated with the cold front had washed the air there was no haze. Visibility must have been about 50 miles. It was a little bumpy down low due to gusty winds, so I kept the 210 climbing in search of smooth air. We found it around 8,000 feet and I leveled off at 8,500 feet. It was a good compromise between the low-level bumps and the strong head wind at higher altitudes.

The Cessna 210 trimmed up nicely and I turned the autopilot on as soon as we leveled off at cruising altitude.

Randy, Ron and I talked softly, while Joyce and Jamie slept. The miles went by slowly, in spite of the retractable Cessna's good cruise speed. Because of the head winds, the ground speed read more like that of a Cessna 172. Although it was very cold outside, the cockpit was warm and cozy. Ron and Randy both like airplanes, so our conversation was primarily centered around the flight, the airplane, the radios and such. They were both full of questions that as a flight instructor I didn't mind answering.

An occasional radio interchange with Air Traffic Control kept me alert to my flying responsibilities. Navigation was a snap. I could see the lights of Birmingham when we were almost a hundred miles away. From there, a slight heading adjustment put us on course for Oxford. The DME kept me posted on our progress and all instruments and gauges on the 210's instrument panel remained well within normal ranges.

When it was time to begin the descent for Oxford, I lowered the nose, picking up speed as we went. When we passed through sixty-five hundred feet, an invisible monster grabbed the Cessna by its wingtips and tried to shake us all out of the airplane.

Immediately Joyce and Jamie were awake. Even in the dull red glow of the cockpit lights, Joyce's face looked as white as a sheet. She never has enjoyed flying much because she has a tendency to get airsick. That has done more to make me a smooth pilot than all the medical evacuation flights I've ever flown. Sensing Joyce's discomfort Randy took the baby out of her lap while I struggled to keep the Cessna upright. *Is this what you call severe turbulence?* I wondered. I'd never experienced anything like it.

I didn't know if it would classify as severe turbulence or not, but I did know I had my hands full and we still had twenty miles and six thousand feet to go. I started thinking of options. Climbing back to altitude wouldn't do any good. We had to come down sometime and our destination was just ahead.

What about an alternate airport? I had no idea what the landing was going to be like, and it was the middle of the night. Trying to go somewhere else didn't make a lot of sense to me right then, especially the way we were already being knocked around. I took a poll and we were all for trying to get the plane on the ground at Oxford.

I wrestled with the plane, while the others tried to stay relaxed. The ride was harder on them than on me. At least I had the control yoke to hold on to. Our seatbelts felt like they could easily cut us in two. We banged our heads on the ceiling during the worst of the bumps. Joyce didn't throw up, but I knew her stomach was protesting. Miraculously Jamie was not crying. Randy was either doing a fine job of keeping him calm or he was too scared to cry.

I slowed to maneuvering speed, and put the landing gear down for more stability. I also added the first ten degrees of flaps. Would coming down faster help? I didn't know, but expected not, so continued the descent at about 500 to 700 feet per minute. There was no sense in adding ear pressure problems to what Joyce and Jamie were already experiencing.

I lined up with runway two-seven at least ten miles out. I couldn't raise anybody on the unicom, but that wasn't surprising since it was after midnight. A check with the Flight Service Station in Memphis let me know their surface winds were out of the north at twenty-five knots with occasional gusts to thirty or thirty-five knots. Memphis was sixty miles away.

After an agonizing final approach that made us all swear off roller coasters and off-road vehicles for life, we were nearing touchdown. I knew it was going to be a crosswind landing, I just didn't know how strong the wind was at the surface. All the way down final I held at least a thirty degree right crab into the north wind. Then it was time to straighten the nose and lower the upwind wing for landing.

That didn't work. No matter how much right wing I tried to hold, we were drifting to the left. The left rudder pedal was against the stop and the nose still wasn't lined up with the runway.

The only thing I had going for me was the fact that I spent four to six hours a day in an airplane, much of it in the traffic pattern. It also helped that I had flown over 800 hours of night combat time in Vietnam. The conditions were tough, but I was fully alert and had a handful of controls at my disposal. If I couldn't get the nose to align with the runway, I'd have to use whatever controls I had to get the plane down and rolling straight.

The arrival resembled a controlled crash. Instead of letting the upwind wheel touch down first, at the last minute I rocked the wings to the left and planted the left wheel. That technique worked! The nose snapped around to the left and immediately I rocked the wings the other way and planted the right wheel. Not a pretty arrival, but we were down and rolling straight ahead. As the airspeed bled off, I fed the control surfaces to the wind, rolling the yoke all the way to the right and pushing the control column forward. There was no way I was going to let that plane become airborne again.

What a wind! When we finally parked and got out of the plane, we had a hard time even standing up in it, much less walking. Perhaps I should have flown to Memphis and landed to the north, into the wind. But, we were down and no harm done.

What did I learn from the experience? I learned that I should have checked the weather a little more closely. Just because the sky was totally clear didn't mean the weather was fit to fly. The wind and turbulence behind a fast-moving cold front made for conditions that probably should have been waited out on the ground.

I learned that when you're in the air and conditions are beyond the limits of the airplane, there is no substitute for hours and hours of practice in controlling the airplane and being aware of your environment and circumstances. And I learned that no matter what the circumstances, never give up. With whatever control you have, FLY THE AIRPLANE!

Cropdusting Fighter Pilots

Date: 3/10/76
A/C Type: Cessna Agwagon
A/C Registration: N4783Q
Route of Flight: Wichita (Cessna) – UOX

It was easy to imagine the I was flying a WWII fighter plane. With its single-seat cockpit, bubble canopy, low wings, and control stick, the plane was powerful for its size and weight. And, it was responsive. I was flying trail in a flight of four. We had been in the air nearly three hours, flying high, above the haze, and with a good tailwind. My bladder was ready for a break and the fuel gauge indicated less than fourth of a tank.

There had been no sign of enemy aircraft, no sign of bombers needing fighter support, just a long stretch of Kansas prairie, blending into the mountains of southern Missouri and northern Arkansas. I hated to admit it, but the novelty of the flight was beginning to wear off. We weren't even flying a tight enough formation to keep the adrenaline flowing.

I couldn't contact lead. Radio silence was in effect, not to conceal our presence from the enemy, but because we had no radios. I hoped lead's fuel and bladder situation were similar to mine and we would be letting down soon.

The sleek, single-seat fighter I was flying wasn't a fighter at all, but a Cessna Agwagon—a cropduster. The mission wasn't a military one, but a simple ferry flight. Still, one can dream.

Ahead I saw lead break to his left in a tight downward spiral. He must have picked out the airport where we could stop for fuel. Two and three broke off behind him, and I followed. The controls stiffened as I dove to catch up with the others, and I could actually imagine myself making a strafing run. Then I saw them—three Grumman Agcats, a couple of thousand feet below us. In tight formation and unsuspecting they were flying straight and level on a southwesterly heading. The element of surprise was definitely in our favor.

What were the odds that a ferry flight of three cropdusters from the Grumman factory in Elmira, New York, would cross paths over northern Arkansas with a flight of four Cessna cropdusters bound from Wichita, Kansas to Houston, Mississippi? It didn't matter! The odds in the impending dogfight were definitely with us. We had them outnumbered and we had the element of surprise!

Lead lined up above and behind the gray and yellow Grumman biplanes. He slowed so as not to overtake them and waited for the rest of us to catch up. It was then that one of the Grumman pilots spotted us and wagged his wings at the others. They broke formation, the two trailing planes turning outside to meet us head on. Man they could turn! We dove right through where their formation had been, and the lead Grumman looped and came around on my tail. I pushed the nose over and began to jink from side to side, but couldn't shake him. He stayed right on my tail and I could hear his imaginary machine guns firing, feel the imaginary bullets raking my fuselage and canopy. I pulled up into a hammerhead and snapped off to the right. He turned the opposite direction, one of those slow, tight turns the biplanes are so good at and as I came out of my dive, he was right there waiting for me, approaching me head on. I was dead. I hoped my wingmen had faired better.

I saluted the Grumman pilot that had bettered me and turned in my seat to locate the others. They were below us, one Grumman diving with a Cessna right on his tail. The other Grumman was caught in a crossfire between the remaining two Cessnas as they strafed him from the side, one pulling up and going over him, the other diving under him at the last minute.

It was over in a matter of minutes. Two Grumman pilots were

dead, but only because they had been surprised and outnumbered. Among the Cessna pilots, there was one dead—me. The Agcats rejoined their formation and continued on their southwesterly heading. We flew by them in loose trail and saluted. The grins on their faces were as big as the grins on ours. Lead took us to the nearest airport to refuel and rearm.

No Step for a Stepper

Date: 4/1/76
A/C Type: Luscombe 8E
Registration #: N2045K
Route of Flight: Oxford – Jackson

I loved Guard drill. I found it more fun than the Army—more professional, too. During the few years after Vietnam I found myself often saying that if I had to go to war again, I'd rather go with a guard unit than with an active Army unit. Everyone in my National Guard unit was a volunteer. They were there because they enjoyed what they were doing. Most of them had been doing their jobs in the Guard for years. From the pilots to the mechanics, to the tank drivers, to the supply clerks, and cooks; they were good at what they did and they had fun doing it.

When I first joined the Mississippi National Guard my unit was in Grenada, but the helicopters were in Jackson. Often I would fly my Luscombe to Guard drill and park it on the Guard ramp just south of runway 29 on the west side of Hawkins Field.

The plan the first weekend in April was to fly the Loach for a couple of hours on Friday evening, spend the night with my dad in Jackson, then fly for the Guard again Saturday and Sunday. Sunday afternoon I would fly the Luscombe back to Oxford. A full weekend of fun flying, and I got paid for it, too!

Heading south Friday afternoon I encountered a strong quartering head wind. As I flew almost sideways down I-55, some of the tractor-trailer rigs below were almost matching my speed.

I enjoyed the time alone in the airplane. It was a beautiful afternoon with puffy white clouds that I poked at with my propeller as I flew along taking in the scenery. The Luscombe is a small airplane, so whenever I was by myself, it was not unusual for me to stretch my legs across to the other side of the cockpit, making it a little more comfortable to fly.

Twenty miles north of Jackson I tuned the radio to Hawkins Tower to listen to the radio traffic. I wanted to get a feel for what was going on down there.

Whoa! What was that? The first thing I heard was a Baron pilot complaining about the low level winds and aborting his approach to runway two-nine. He said he was going over to Thompson Field to land.

For a few minutes the traffic pattern was quiet. Not much traffic. Could it have been because of the wind?

Then a Twin Beech reported in on final. I listened carefully to the Beech pilot's dialogue with the tower controller.

"Say winds."

"Two-forty at twenty-six knots, gusts to forty."

"Feels it."

"I can imagine."

The conversation was not very reassuring considering he was flying a 10,000 pound freighter and I was flying a little Luscombe that barely weighed 2,000 pounds full of fuel and with me in it. I grew uneasy and called the tower. Might as well commit. I had to land somewhere. "Hawkins Tower, Luscombe two-zero-four-five-kilo, fifteen miles north for landing."

"Two-zero-four-five-kilo, Hawkins Tower, winds two-four-zero at twenty-five, gusts to forty, altimeter three-zero-zero-two. Plan a right base entry for runway two-nine."

"Four-five-kilo. Pretty good wind down there," I commented.

"No step for a stepper."

No step for a stepper. What did he mean by that? Did he think that just because I was flying a thirty-year-old airplane that I had thirty years' flying experience and could handle anything thrown at me? Not so!

The remaining distance to the airport seemed to take an hour. It really wasn't that long, but the miles sure seemed to go by slowly with me dreading the landing. I would like to tell you that this professional pilot/flight instructor was not worried, but I'd be lying. In fact I wished I was somewhere else. The Luscombe was notorious for being hard to handle on the ground. I hadn't ground looped it yet, but I had an ominous feeling about the winds at Hawkins Airport that afternoon.

It seemed to take forever because of the wind, but I finally arrived on an extended right base leg for the runway and reported in. "Hawkins Tower, Luscombe four-five-kilo, right base for landing."

"Four-five-kilo, cleared to land. Winds now two-six zero at thirty knots."

"Four-five-kilo, roger, cleared to land." I answered and hung the microphone on its little clip on the instrument panel. Now the work began. Winds thirty degrees off the runway heading wasn't bad. Thirty knots was bad.

I turned final and just hung in the air. The airspeed indicator said fifty knots, but I swear the Luscombe seemed to be flying backwards. The approach began to resemble that of a helicopter. At least that was familiar territory.

It took considerable power to make it to the runway. The engine and propeller pulled me slowly toward the airport. I set up the proper glide angle and waited.

I didn't feel any gusts, just that steady wind thirty degrees off the nose. I left the Luscombe's nose pointed into the wind and the ground track carried me to the runway. I figured to pop the nose around with the rudder at the last minute, lowering the left wing to keep the plane headed straight.

When the plane touched down I was amazed. The Luscombe simply settled onto the big, white number 29 painted on the end of the runway and stopped. No ground roll, no fighting the crosswind, just a kissing of the pavement. I closed the throttle and we stopped dead in our tracks. Wow! That was cool.

"Four-five-kilo, where are you parking?"

"At the Guard ramp."

"Taxi to the Guard ramp. You can remain on this frequency."

There were no helicopter flight training periods that night. Too windy.

LOHs to North Carolina

Date: 5/15/76
A/C Type: OH-6
Registration #: 12956
Route of Flight: M42 – Golden Triangle – Anniston – Anderson –
Lumberton

Summer camp this year was going to be in North Carolina. We picked up our helicopters in Jackson on May 14th. That night we flew up to Grenada to get our orders and to pick up our crew chiefs. We left Grenada the next morning, four OH-6s flying in formation, destination—Lumberson, North Carolina. Our first fuel stop was at the Golden Triangle Airport in Columbus. From there we flew to Anniston, Alabama, then to Anderson, South Carolina.

The further east we got, the worse the weather was, with low ceilings and rain. Since I had flown in North Carolina for three years, the others had decided I would fly lead. As we began flying over the lower portion of the Smoky Mountains, I found myself picking my way through mountain valleys with low ceilings and reduced visibility due to the rain. It was no good and getting worse. I radioed the others that I was turning around and started a turn to the left. One of the guys behind commented that it was a good thing I had turned around because they were about done following me if I hadn't.

We refueled in Anderson, South Carolina, then took up a course to Lumberton, North Carolina. Lumberton was familiar territory to me. I had flown both my airplane instrument check ride and my multi-engine check ride at the Lumberton Airport, plus it had been the site of much of my helicopter instrument training. We were flying at 300 feet altitude with about a half to a quarter mile visibility. I kept the airspeed around fifty knots. Navigation was pure dead reckoning as there wasn't enough visibility to pick out landmarks. Lumberton had a non-directional beacon and I had it tuned in, hoping the heading we were flying would get us close enough to pick it up.

The flight from Anderson to Lumberton was two hours long. It was late afternoon and a steady rain was falling. It was taxing work flying under those conditions, keeping a constant vigil for radio towers and hoping you weren't way off course and would miss the destination entirely. Our helicopters were like cocoons, keeping us from the weather.

The other three LOHs were following me in loose trail formation, apparently confident that I could find Lumberton. I wasn't that confident myself, but when we crossed I-95 the ADF needle came alive, and I knew we were close. I followed the needle right to the Lumberton Airport and landed on the ramp. There were some guys from our unit already there waiting for us. They had driven over in a convoy a couple of days earlier. They seemed in a hurry as they began putting ground handling wheels on the helicopters to put them in a hangar. That's when I found out the whole area was under a tornado watch. We had talked with Flight Service and had received several FAA weather briefings during the flight over and not one time had anyone mentioned tornadoes!

That night, a tornado touched down at the airport, damaging a few buildings, but thankfully not the one that housed our OH-6s.

Another Taildragger

We outgrew the Luscombe, since we now had our second son, Nathan, and began to look for a four seat airplane. I was in love with the idea of having a taildragger. It seemed like real flying to me. A Cessna 180 or 185 would have been the dream plane at the time, but a Cessna 170 was closer to the realities of the checkbook.

The way we found N2702C was one of those "divine appointment" kind of things. It was just meant to be that we would have that particular airplane. Mike McKee and I had flown a charter to Kentucky for Holly Carburetor and on the way back we both needed a potty break. It was about 10:00 at night when we landed at Lexington, Tennessee, for no other reason than the fact that it was the closest airport.

As we taxied over to the ramp, we fell in love. Well, I did. Mike was the level-headed one among us, as I would soon find out when we started negotiating. There she was, a beautiful baby blue and white Cessna 170 with a *For Sale* sign stuck in her window. In spite of the hour, the airport manager had seen us land and had driven out to see if we needed any fuel and when he saw us eyeballing the 170 he called the owner, who promptly came out to the field to answer questions.

No, the price was not negotiable. "Come on let's go," Mike said. I wanted to slug him. No price was too high to ask for this beauty. I wanted it and I wanted it now. Fortunately, Mike's tactics worked. We left that night without having our offer accepted, but we did make a deal the next day for $10,000 even, $800 less than the asking price.

If you have never flown a Cessna 170B, you've missed something. The sheer joy of being behind the controls when that airplane takes to the air gives a pilot a sense that he is part of something beyond just flying. This is an airplane that leaps into the air with glee, that settles into trim like she's on a greased rail, that purrs through the air all mellow and serene and occasionally lets you stroke her neck or mane as she eagerly takes you where you want to go. Huge flaps, smooth controls, delicate lines, it's a graceful machine and I regret that I don't still own her today.

Mike McKee and I bought 02C together and flew it around North Mississippi for the better part of a year. My friend Randy Peters and I built a modern instrument panel with up-to-date gauges. Mike was in the radio business, so he upgraded the avionics to the latest and greatest. The plane had a belt-driven vacuum pump instead of the venturi tube that stock 170s had for supplying suction to the gyros. We did all the checks and had the plane certified for IFR flight.

Calling Flight Service

Automated Flight Service Stations and the current 1-800-WX-BRIEF phone number that works anywhere in the US is a relatively recent development. During the time I was flying regularly in Mississippi, there were Flight Service Stations located at many airports throughout the country. A local Oxford telephone number (234-3434, if I recall correctly) rang in the Memphis Flight Service Station, which was how we normally got our weather briefings and filed our flight plans. Flight Service Stations were also located in Greenwood, Tupelo, and Jackson.

One day I dialed the local FSS number and the briefer answered the phone, "Flight Service, Greenwood."

"Greenwood? I was calling Memphis!" I stated, not understanding how the local number for the Memphis FSS could have been routed to the Greenwood FSS.

There was a chuckle at the other end of the line. "This is Memphis," the briefer stated. "My *name* is Greenwood!"

Airborne and Now What?

Date: 3/13/77
A/C Type: Cessna 170B
Registration #: N2702C
Route of Flight: Tupelo – Oxford

There were better opportunities for me in the National Guard unit in Tupelo since they had both Hueys and OH-6s. I had been a Huey IP for several years, now I was an OH-6 IP as well. So, I transferred to the unit in Tupelo. That's why I was sitting in a classroom in the armory in Tupelo late on a Sunday afternoon when I got called to the phone.

"Mr. Freeman, this is Jim Holley over at the FBO office. Are you the one flying the blue and white Cessna 170?"

"Yes, I am," I answered.

"You'd better get over here and find some way to tie it down. The wind has picked up considerably and that plane keeps trying to fly by itself."

I located my unit commander in the back of the classroom and told him I needed to go tend to my airplane. As a pilot himself, he understood and quickly nodded an assent. I grabbed my flight jacket and headed out the door.

It was a three hundred yard walk from the armory to the FBO and I was bracing a stiff south wind all the way. That was a switch— strong winds right out of the south.

The 170 was dancing in the wind when I arrived at the ramp. There had been no tiedowns available when I landed that morning, so I had chocked the wheels and left the plane, not realizing the wind was going to pick up that much during the day.

The plane was weather-vaned into the wind, having jumped one of the chocks with its right wheel. It was a minor miracle that the wing tips hadn't hit another plane.

I wasn't sure what to do. I looked around for something to use as a tiedown and didn't see anything. I had to do something to protect my baby, so I climbed into the cockpit and applied the brakes. The wings were almost flying; I could feel the wheels trying to leave the ground.

I fished in my pockets for the key and started the engine. That gave me some control. Turning on the radio, I called Flight Service.

"Tupelo radio, Cessna two-seven-zero-two-charlie, requesting airport advisory."

"Cessna two-seven-zero-two-charlie, this is Tupelo radio. Winds out of the south at 30 knots, gusting to 45, no reported traffic."

I didn't know what else to do but take off. I couldn't sit in the plane all night with the engine running and I knew that the one hangar on the airport was full. I figured perhaps conditions at Oxford weren't as bad.

I took off to the south. The ground roll was almost non-existent.

During the thirty-minute flight to Oxford, I analyzed my options. The plane was full of fuel. I could try to land at Oxford. If it was too windy, I could go over to Holly Springs, which had a north-south runway, or up to Memphis. I could also go back to Tupelo, but I still wouldn't have a safe place to park the plane.

Jeff Meaders, the airport manager at Oxford, answered the radio when I called for an airport advisory.

"Good afternoon, Cessna zero-two-charlie," he greeted me. "Winds are out of the south, steady at thirty knots. No reported traffic. Runway of your choice."

"Zero-two-charlie," I acknowledged. Of course there was no reported traffic. No one else was stupid enough to be flying in that wind.

The runway at Oxford runs due east and west. That meant a direct thirty-knot crosswind. Obviously, this was too much for the taildragger Cessna. I considered landing on the taxiway between the runway and the ramp. It was short, but it ran north and south and I could land directly into the wind, and would have very little ground roll after touchdown.

But, at the end of the taxiway there was a fuel shack, and if I didn't get stopped in time to avoid the shack, I'd have two options: hit one of the parked airplanes on one side of it, or go to the other side and head down the big downhill slope behind the shack. It seemed to me the risks outweighed the advantages. Still, it was an option I mulled over.

I was definitely not going to land on runway two-seven. There is an area of rough terrain—hills and gullies—just off the approach end of runway two-seven that causes considerable turbulence on windy days. I entered the traffic pattern for runway nine. I decided to try a wheel landing. I planned to keep the speed up and maintain rudder control as long as possible. I also planned to land on the grass beside the runway. I often practiced wheel landings there with my students. For some reason, I found it easier to do good wheel landings on the grass than on the pavement.

Jeff and I had several conversations about the wind and my options for landing. I learned during those conversations that Jeff wasn't the only one at the airport. Several of my students were also there and would be outside watching as I attempted to land. Great!

Landing on the grass meant I would land long, touching down past the taxiway. I turned final lined up with the landing area. Jeff gave me a last minute update on the wind, then went outside to watch. It was still blowing directly across the runway at thirty knots.

I made a shallow approach. As I went by the terminal building, I could see them all standing by the fence watching. Six or seven guys were about to see their flight instructor bungle a landing—big time.

The plane neared the grass and I didn't experience any alignment problems. In fact, I didn't experience any problems at all. The nose lined up just fine. I even had to level the wings to keep from veering off to the right—into the wind. Strange.

No it wasn't. It was the trees! The grass landing area was bounded on the south by a thick stand of mature pine trees and the trees blocked the wind. I made one of the most beautiful wheel landings I had ever made. I could almost hear Jeff's and the student's applause (I wonder if they really did). The trees totally negated the effects of the wind. Should I tell them? Na-a-a-h.

Second UFO

Date: 6/10/77
A/C Type: OH-6
Registration #: 68-17175
Route of Flight: Tupelo – Oxford

On a Friday night after work, Mike McKee, my friend and partner in the Cessna 170, flew me over to Tupelo to pick up a helicopter for the weekend. The 170 was in the shop having the engine overhauled, so we flew over in a rented 172.

I preflighted the helicopter and we left Tupelo just after dark, me flying formation off Mike's right wing. We flew over the high school football stadium in Pontotoc where a football game was in progress, and headed toward Oxford. When we were about fifteen miles east of town, a huge ball of flame fell out of the sky in front of us. "Did you see that?" Mike radioed.

I answered, "I sure did, what was that?"

He didn't know. It appeared that the ball of fire had hit the ground eight or ten miles ahead of us. We flew to the area where we thought it had gone down, believing that we would find a fire on the ground. We circled around a bit, but saw nothing.

"Mike, David, this is Jeff. Are you guys all right?" It was Jeff Meaders, the airport manager, calling on the Unicom frequency to try to reach us. I was quite surprised at his call, and answered him right away.

"Yeah, Jeff, we're all right. Why?"

"We got a call from the police a few minutes ago. A trucker reported seeing an airplane on fire falling out of the sky. I just wanted to make sure it wasn't one of you guys."

"We saw it, too," I radioed back. "It was an airplane? We couldn't tell."

"That's what this guy said," Jeff responded. "Said its wings were on fire."

Since we couldn't see anything resembling a plane crash, we both flew over to the airport and landed. Jeff met us on the ramp and told us that the Civil Defense and the Sheriff's Department were out searching.

"Mike, I don't think what we saw was an airplane."

"Me, either," Mike replied. "My guess is that it was a meteorite."

"A meteorite . . . now that makes sense," I agreed.

We refueled the helicopter, Mike got in with me and we flew back out to the area where the guys on the ground were searching. They were very close to where we thought we'd seen the fireball go down, but they weren't finding anything. They reported a little activity on some Geiger counters, but that was it.

The next morning early, I was scheduled to go to Jackson in the helicopter. I flew over the area in the daylight, expecting to see a large burned area or something, but saw nothing. In the week or so that followed, Mike and I were both interviewed by some UFO chasers, but as far as I know, no one ever identified what it was we saw that night.

Chapter 5 – Corporate Pilot

Flying Computer Repairman

We moved to Texas in the summer of 1977, where I was fortunate enough to find a flying job. A small, but rapidly-growing company needed a multi-engine pilot. The guy who interviewed me was not a multi-engine pilot and he and I hit it off so well that he either forgot to ask me how many multi-engine hours I had, or didn't know to ask me. He didn't even explain that pilots at that company filled a dual role. Being a pilot was secondary. The primary job was as a computer repairman. I found that out after I reported to work.

I told them that I didn't even know how to use a computer, much less repair one. "That's okay," they said, "we'll teach you."

I was scheduled for a two-week training class. Halfway into the second day, the manager of the service department stuck his head in the door and asked me if I knew how to fly a Cessna 210. I told him I did, so he said to come with him. That afternoon we flew to Abilene, Texas, on a computer service call. I never did make it back to the classroom.

The company that I had begun working for was V. C. Brown, Inc. It was the first company in the US to begin installing standalone computer systems in pharmacies. If you get prescriptions filled, you know that pharmacies maintain prescription records, print labels, etc., on computers. This was the early days of that business and V. C. Brown was a pioneer. He was also a pilot and had come up with the concept of hiring young guys that knew how to fly and training them as service technicians. That way instead of having regional or district offices all over the country, the company could remain centralized and its pilots could fly out to small town America to take care of service problems. Using private airplanes for this task in those days made sense because most of the pharmacies were not in large cities, and because the computers of that day were big. The spare parts that it took to keep them running were not exactly carry-on luggage for the airlines.

I had this job for three years and it was fascinating work. Over time we came to have a whole fleet of airplanes and a bunch of pilots, and I became the Chief Pilot, the daddy of them all. We flew to places like Fort Scott, Kansas; or Poplar Bluff, Missouri; Salem, Arkansas; or Montrose, Colorado, amassing lots of time in the air, and much of it in actual weather and/or at night.

When I went to work in the morning, I rarely knew where I would be going that day or when I would return. Joyce and I made a game out of it. I'd call her and say, guess where I am and she'd guess something like Tampa or Odessa or Kansas City. She got it right pretty often because there were "problem" customers in certain areas that required a lot of service calls.

Joyce would often have to throw a bag together for me at a moment's notice. She would just ask, "how many days?" I would run home, get the bag, kiss her and the boys goodbye, then go to the airport and take off.

One night after such a quick takeoff and a long day's work at a pharmacy, I had just taken a long, hot bath at the motel. When I got out of the tub and pulled out the clean underwear Joyce had packed for me, it was a size 4, little boy's pair belong to Jamie. Needless to say I washed out my dirty pair and hung them up to dry during the night. We have laughed over that night many times since then.

It was on this job that I began to accumulate multi-engine time. These were the days before the insurance companies required so many hours of multi-engine time, so many hours "in-type," so many hours of total time, formal schooling on the aircraft and your first-born son or daughter before they would cover you. The company owned a Cessna 402 when I came on board and later traded that plane on a Cessna 421. I don't know how they got me on the insurance policy; I wasn't involved in those details, but I started right out flying the Cessna 402, because there weren't any other multi-engine pilots working there at the time, except for the boss and he couldn't make all of the flights.

When we moved to Texas, Mike McKee graciously sold his part of our Cessna 170, N2702C to my former Luscombe partners, all of whom were moving to Texas about the same time I was. I made many flights back and forth from Oxford to the DFW area flying men who were seeking jobs and their wives who were looking for houses, during the time my friends and I were relocating to the Dallas-Fort Worth Metroplex.

After moving, I flew 02C some for my new employer, but that fizzled out when he bought more company planes. So she sat. First at Mangum field in Hurst, then when they closed that field, she sat at Grand Prairie. She sat because I was flying in excess of 120 hours a month in my new job.

Then I got a call from the man who had originally sold us the airplane. He wanted it back and wanted to know how much we would take for it. I talked with my partners and they all wanted out. I quoted a price that was what we had bought the airplane for plus what we had put into improvements and the man came and took her away. I didn't feel the loss at first, but later, after my other flying began to taper off, I looked around and realized that selling 02C had left a big void. We tried to fill it for a while with a Cherokee Six, but it just wasn't the same.

Many years had passed, at least eighteen, maybe twenty, and I learned how to look up ownership of airplanes on the Internet. I looked up N2702C and discovered she still belonged to the guy we had sold her to. I called him. He agreed to send me some pictures, but selling the airplane was out of the question—at any price. If you're ever fortunate enough to own a Cessna 170B, especially one with a full set of IFR gauges and radios and a vacuum source to drive them, don't ever sell it. Ever!

Multi-Engine Pilot

Date: 9/7/77
A/C Type: Cessna 402B
Registration #: N98764
Route of Flight: Fort Worth – Montgomery – Fort Worth

My second day on the job, the boss and owner of the company, V. C. Brown, checked me out in his Cessna 402. We took off from Meacham Airport in Fort Worth and flew up to Denton to do an instrument approach and some takeoffs and landings. I'd never flown a Cessna 402, but had made a couple of flights in the right seat of a Cessna 310, which was somewhat similar.

I guess the boss was satisfied, because the next morning, which was my third day on the job, he sent me in the 402 to Montgomery, Alabama, with two computer systems and their trainer/installers.

On the way over I flew smoothly and carefully. On the way back I was by myself, so I let the autopilot fly while I read the POH. Being young and having spent several years as both a military and civilian flight instructor, I was able to adapt easily. Over the next few months I began learning to use radar and deicing equipment. When I was by myself, I practiced engine-out procedures and other emergencies and soon became quite comfortable with the airplane.

I'm sure I missed a lot of the basics because I did this all on my own without the benefit of an older and more experienced pilot as a mentor. When we later bought a pressurized Cessna 421, my get-acquainted flight in that plane was an actual flight that involved bad weather, mountains, and using some new equipment that I had never used before, such as a horizontal situation indicator (HSI) and RNAV area navigation equipment. Yes, I know this stuff is old hat now and some of it even outdated, but in 1978, it was all very new.

The Perfect Instrument Flight

Date: 11/23/77
A/C Type: Cessna 402B
Registration #: N98764
Route of Flight: Fort Worth – Little Rock – Fort Worth

A potential customer in Little Rock had indicated that he was going to buy not one, but three computer systems. It was an appointment the National Sales Manager felt he had to make, even though the weather was IFR with low ceilings on both ends. I filed for 9,000 feet with Little Rock as my destination and Memphis as the alternate.

The IFR release came from Meacham Tower, throttles up, V1, rotate, gear up, in the clouds, just about that fast. Departure held us at 4,000 feet straight across DFW, which is the wrong altitude for that direction, but was the altitude they always assigned eastbound traffic across DFW. Then came the climb to nine and the switch to Fort Worth Center and all the time we're in the soup. Often on overcast days you can climb a few thousand feet and be in bright sunshine above the clouds, but not this time. I didn't even find a place between layers. Fortunately there was no ice.

It was solid IFR, and I was on the gauges all the way to Little Rock, where I flew an ILS approach to minimums and there was the runway, right where it was supposed to be, easily spotted through the half-mile visibility under a 200-foot ceiling compliments of the high intensity runway lighting.

Three hours later the trip back to Fort Worth was the same way— solid instruments all the way, ending in an ILS approach to minimums. The sales manager, Bob, was really impressed with my flying. He slept the entire way back. How, I wondered, could a person sleep through such an exceptional flight compliments of air traffic control and a well-equipped airplane with a coupled autopilot?

When we touched down on the runway at Fort Worth I flipped the electric fuel pump switches from the "high" position they had been in for landing to the "low" position needed to keep the idling engines running on the ground. At least that's what I meant to do, but on the right engine, I didn't pull the switch all the way back like I was supposed to and it stopped in the middle or "off" position. Since the right engine was no longer getting the proper amount of fuel it stopped while we were rolling down the runway. When we reached the taxiway and it came time to make a left turn off the runway to go to our hangar, I couldn't do it. Applying power to the left engine made the airplane turn to the right and there was no power from the right engine to counteract the thrust. I couldn't overcome the power from the left engine with rudder and nose wheel steering alone, so I had to make a 270 degree turn to the right to get the airplane headed in the right direction.

Bob woke up when he heard me goosing the engine, looked out his side of the airplane and saw the right propeller stopped dead still.

"How long has that engine been off?" he asked.

"Since right after we took off at Little Rock," I casually replied with a straight face. I think he believed me.

Southern Pilot in the North Country

Date: 12/12/77
A/C Type: Cessna 402B
Registration #: N98764
Route of Flight: Fort Worth – Evansville – Dayton – Chicago – Springfield
– Fort Worth

The sales staff sold in the south all summer, but chose the winter months to make sales up north. Being a southern boy who had learned to fly in the south and had done most of my flying in the south, I had some interesting experiences when I started flying to the upper midwestern states in the winters of 1977 and 1978.

I had only flown in snow once before in my life and that was during that ferry trip across the south in a helicopter and the snow wasn't very heavy. Now here I was flying into Dayton, Ohio, in heavy snow and hearing on the ATIS that braking action for the active runway was nil. I didn't have a clue what that meant to me, but since DC-9s and 727s were landing on that runway I figured it must be all right for a Cessna 402. As it turned out it was no problem at all.

Leaving Dayton after dark, I flew up to Chicago's Midway Airport. It was still snowing. With the lights off, you couldn't even tell it was snowing, but turn on the strobes or the landing lights and it was all white stuff streaking through the sky. It was pretty and not much of a bother because none of it was sticking to the airplane. Over the southern tip of Lake Michigan we descended out of the clouds and below, as far as I could see, the yellow street lights of Chicago were laid out in rows, east and west, north and south. The southern tip of the lake was frozen over and covered with snow. On the ground at Midway snowdrifts six feet high lined the runways and taxiways. I had to be careful taxiing because I couldn't see around the corners.

During these winter flights up north the deicing boots on the wings and tail of the airplane served me well, as did the heated windshield and propellers. It was on a night like this that I experienced my first St. Elmo's fire dancing around the windshield—a blue electrical glow that snapped and popped as it dissipated the static buildups from the heated Plexiglas. The first time I heard ice being flung from the propellers against the side of the fuselage, I thought the airplane was coming apart.

I learned to have the plane put into a hangar for overnight storage whenever I could, so it would start in the morning. More than once at a small airport I had to shut down on the taxiway and have the airplane towed to the ramp to keep from hitting piled up ice and snow with the props, which would have certainly bent them into strange angles and caused sudden stoppage in the engines, necessitating a tear-down of the engines to inspect for damages.

One of our guys ran off the end of the runway at Grandpa's Farm Airport in Mendota, Illinois, because he couldn't get the 210 stopped in time and another guy struck ice with a prop at Centralia, Illinois. Several nights I landed at Midway and taxied to a ramp filled with Twin Beeches waiting to take their night freight runs as soon as the ice could be cleared from their wings and from the taxiways. The airport at Springfield, Illinois, became a favorite overnight spot because there was a motel located at the airport adjacent to the transient aircraft parking area.

The Lonely Cockpit

Date: 12/22/77
A/C Type: Cessna 210
Registration #: N2255S
Route of Flight: Poplar Bluff – Fort Worth

It was around 11:00 p.m. when I finally arrived back at the Poplar Bluff Airport for the return trip to Fort Worth. SEMO Airline's DeHaviland Huron was sitting on the ramp, quiet and silent. All of the airport personnel had gone home for the night. All I knew about the weather was that it was cold and when I looked up I saw clouds. They looked close.

I found a pay phone and called Flight Service. The nearest reporting station was Cape Girardeau, which had 200 and 1. Memphis was about the same. Fort Worth Meacham, DFW, and Dallas Love were all clear with visibility unlimited. I filed IFR, 2.5 hours enroute, direct Little Rock, then the airways to Blue Ridge, direct Meacham, knowing I'd get the Blue Ridge arrival from DFW Approach Control.

The takeoff was uneventful, but I was on the gauges right away and had to turn off the strobes because of their reflection off the clouds. Level at six thousand the 210 trimmed out nicely. I switched on the autopilot and sat back to enjoy the ride.

It occurred to me a few minutes into the flight that the radio was quiet—too quiet. I began to wonder if it had failed. I called Center.

"Memphis Center, Cessna two-two-five-five-sierra, am I the only one up here tonight, or are you just working low altitude?"

"Five-five-sierra, Memphis Center, I'm working all altitudes, you're just the only one I have tonight."

At least the radio was working, but immediately I began to wonder what everybody else knew that I didn't know.

"Unusually light traffic, isn't it?" I queried the controller.

"Well, Memphis is below minimums," he answered. "So is St. Louis. Chicago has been closed all day."

This didn't sound good. "What about Little Rock?" I asked.

"Little Rock is below minimums, too."

"And Texarkana?"

"Texarkana is zero-zero."

That part wasn't so unusual. Texarkana had been zero-zero a lot lately. But with all the airports being closed I didn't have any place to go if something went wrong. Why wasn't this brought out in my weather briefing? Could it have been that I didn't ask the right questions? I was really beginning to feel uneasy, but had to ask just one more question, just to make sure I hadn't missed something.

"What about Dallas?" I asked.

"Dallas is clear and visibility unlimited," Memphis Center replied.

So, I was okay after all. This was a Mississippi Valley thing and I was headed west. I relaxed.

A few minutes passed and my imagination quit making up imaginary headlines about a certain Texas pilot dying in a plane crash, possibly weather-related.

Out of the blue Memphis Center asked me a question. "Five-five-sierra, are you getting any ice there at six thousand?"

ICE! I hadn't even been thinking about it. I should have been thinking about it, but I wasn't. I grabbed the flashlight from the pocket behind my seat and shined it out at the wings. All I saw was gray murk. "I don't know," I told the controller. "It's too dark to tell."

"Roger." Such a simple, one word response. Somehow it seemed so final. Like, "You're stupid, and you're probably going to die because of it."

I flew along, imagining inches of ice building up on the upper surfaces of my wings, just out of sight. Yet the 210 continued to cruise normally, as smooth a flight as I could remember. Somewhere in western Arkansas I was handed off to Fort Worth Center. The controller there confirmed for me that Fort Worth Meacham was still clear with visibility unlimited. He didn't bother me with any questions about ice.

Near Paris I flew out of the clouds. The lights of the DFW Metroplex were visible on the horizon. People were actually talking on the radio. Not many, as it was now close to 1:00 a.m., but at least I no longer had the feeling I was the only one flying that night.

Such a simple flight, well within the capabilities of both pilot and plane. It was a routine flight, a businessman returning home after a productive day of work. A beautiful flight, actually. But for a while there it was quite lonely in that cockpit. Zero-zero, indeed!

Night IFR Single Engine Failure

Date: 12/30/77
A/C Type: Cessna 210
Registration #: N2255S
Route of Flight: New Orleans – Fort Worth

N2255S and I had become old friends. The week after Christmas I flew it to North Carolina to make some system deliveries and then back to Texas. On December 30 I flew down to New Orleans in 55S. The trip took one hour and forty-five minutes. The weather was IFR and I was between cloud layers most of the way.

The first of two service calls in New Orleans was successful, but on the second one, it became obvious that I was going to have to take the computer system back to Fort Worth for repair.

It was after dark when I left New Orleans. I filed for 8,000 feet. Within a few minutes after departure I was cruising along serenely between cloud layers. The layer of clouds below me was thin enough that I could see the lights of various towns as I flew along.

New Orleans Approach Control handed me off to Baton Rouge Approach. Suddenly the engine surged a few times, then quit. I raced through the engine out emergency procedures: Electric fuel pump on high, mixture full rich, fuel on the fullest tank, and was somewhat successful in getting some life out of the engine. It would run at a power setting below eighteen inches of manifold pressure, but above that it would start coughing and sputtering and threatening to quit.

I declared an emergency. Baton Rouge wanted to know my intentions while telling me that the weather at Baton Rouge was 400 feet overcast with two miles visibility.

I could tell I was just east of the city because of the reflection of the lights below, but I couldn't tell where the airport was. I trimmed the airplane for best glide speed and asked the controller to vector me to a position over the airport. I told him to line me up with the ILS localizer when I was down to 2,000 feet.

I was calm and in control as the airplane descended. Seeing the lights through the clouds gave me confidence. The controller gave me a position report and I realized that he was vectoring me out toward the outer marker like he would in a normal ILS approach. I didn't have enough altitude for that, so I quickly told him that wasn't an option. "Turn me toward the localizer now," I told him. "I'll get down."

He gave me a heading that would take me directly from my present position to the localizer. It was close. I intercepted the localizer just about the same time I broke through 400 feet and there was the runway right in front of me. I made a smooth landing for the benefit of all of the firefighting and rescue personnel that were lined up along the runway waiting for me. I expressed my thanks to them over the radio for being there. The tower arranged to have a tug come out and tow me off the runway.

The next morning a local mechanic checked the airplane over and couldn't find anything wrong with it. He did a full power runup and it seemed to run just fine. I got in and taxied the plane around the airport and it seemed to run okay. I took off and flew a close-in traffic pattern. Still no problem. The engine failure from the night before was a mystery.

Finally, since it was VFR, I took off and headed up toward Alexandria, keeping a big highway underneath me. I wasn't two miles from the airport when the engine started surging again. I quickly turned around and managed to get the plane on the ground before the engine quit completely. This time the mechanic found the problem and replaced both fuel pumps, the engine-driven one and the electrical one. After that, the trip home was without incident.

All pilots train and prepare for an engine failure. But we all hope that when and if one happens to us, it won't be at night or in instrument conditions, or in a single-engine plane. Mine met all three criteria. I figure the odds are against it ever happening again.

Green Clouds Over Gregg County

Date: 6/7/78
A/C Type: Cessna 206
Registration #: N7339C
Route of Flight: Fort Worth – Longview – Shreveport – New Orleans

The majority of our V. C. Brown service call flights were flown solo. On the day of this flight, however, there were several customers in southern Louisiana with down computer systems. Jim Gibson, who I had flown with in the Army at Fort Bragg and who was now also flying for V. C. Brown, and I loaded up a Cessna 206 with parts and headed off for New Orleans to take care of them as quickly as possible.

The forecast mentioned the possiblity of thunderstorms along our route of flight. That was standard fare for a Fort Worth to New Orleans flight in the late spring and early summer. You never could tell what would actually be there until you went to take a look. It was rare that a pilot couldn't find a way through or around any buildups with the help of Center and Approach radar controllers and a good set of eyeballs. This was long before the days of Stormscopes and radar in single-engine airplanes.

We were approaching the Gregg County VOR in east Texas and things were looking pretty dark ahead. I was flying; Jim was in the right seat and talking with Longview Approach. Our flight planned course was along the Victor airway to GGG (Greg County VOR), then turn right to follow the airway that went over Alexandria, Baton Rouge, to New Orleans.

"It doesn't look so good," Jim said.

"Sure doesn't," I agreed. "Ask the controller if he sees a way through this stuff."

He did and the controller came back with a suggested heading of 150 degrees. That was the way we wanted to go and it did look a little lighter in that direction than any other. So far we had been clear of clouds, but we were just about to go into the soup, losing all contact with the ground. Jim wanted to turn around. He was by nature more cautious than I. He also wasn't the one that would get screamed at by angry customers when they found out help wasn't coming. I was. I wanted to at least try it.

Cautiously we entered the clouds.

"I don't like this," Jim said. I was feeling a little jittery myself. Suddenly, we both exclaimed at the same time, "These clouds are green!"

Green clouds are bad news. I wouldn't have ventured into green clouds, but here we were. They had been white, almost gray a few seconds ago, but now the clouds all around us had a greenish tint to them. It was a menacing color, one we had learned to associate with heavy rain and heavy turbulence. I started a left turn as Jim keyed the microphone to advise Longview Approach we were turning around.

As I banked left I happened to look out my side window and there right below me was a runway. All I could see were the numbers "13" indicating the approach end of runway 13 at the Gregg County Airport. "Tell him we have the runway in sight and want a contact approach," I told Jim. He advised what sounded like a very skeptical controller and we were given a clearance for a contact approach to runway 13 at Gregg County.

We were at 7,000 feet. That runway was a long way below us and the hole I was looking at it through was a small one. But we were helicopter pilots and knew how to make a steep approach. The 206, when you chop the power, shove the propeller control all the way forward to put the prop blades in flat pitch and put down all 40 degrees of flaps, almost becomes a helicopter. It will descend steeply and you can easily bleed off speed by holding the nose up a little.

Down we came, spiraling to stay in that hole and keep the runway in sight. I blessed the big Cessna single for its flexibility and maneuverability and asked of it bush pilot performance as we got it on the ground.

The landing was right on the numbers. Visibility on the ground was pretty low, less than half a mile. By the time we had taxied to the ramp it was beginning to rain. We jumped out and headed for the FBO. A twin-engine turbo-prop roared overhead. When we were inside, we learned what we had heard was a King Air executing a missed approach. Weather was now below minimums.

In the pilot's lounge a radio monitor tuned to the approach control frequency allowed us to listen to inbound traffic. Within the next fifteen to twenty minutes after we landed two more airplanes missed their approaches. Our getting in had been a fluke.

We waited until the worst of the storms had subsided, checked the weather and decided we could make it on to New Orleans. We were wrong. After a short flight we caught up with the menacing line of storms again and landed in Shreveport to wait them out. We finally made it into New Orleans late that afternoon.

We're Going Where He's Going

Date: 6/22/78
A/C Type: Cessna 210
Registration #: N2255S
Route of Flight: Atlanta – New Orleans

It was late in the afternoon when I was handed off to New Orleans Approach for sequencing into Moissant International. Cumulus buildups were in plentiful supply, so even though it was bumpy down low, I was grateful for the lower altitude.

Approach was working quite a few airplanes. Landing was to the south and we were being put in the daisy chain north of Lake Pontchartrain. I was following a Texas International DC-9 and there were other aircraft following me. I'm sure the slower speed of the Cessna 210 was causing the sequencing controller to have to work a little harder, but I was keeping my speed up as best I could.

Suddenly, the DC-9 pilot ahead of me made a very unprofessional radio call. "We're getting the hell out of here!" he said.

"Say again?"

"This is TI three-nineteen and we're getting the hell out of here!"

"Roger, TI three-nineteen," said the controller, "what are your intentions?"

"We've got lightning all around us," the TI captain replied, "and our intentions are to get the hell out of here. We're turning right 90 degrees."

"Roger." The controller then turned his attention to me. "Five-five-sierra, what are your intentions?"

"I want to follow him," I said. He had radar, I didn't.

"Okay, five-five-sierra, turn right to two-seven-zero for now."

The TI captain had apparently flown out of the worst of it, but he wasn't going to turn south again. He began asking the controller for help getting around the weather. The controller's response wasn't too assuring.

"I'm picking up cells all over my scope right now," he said. "They've just started popping up within the last five minutes."

It was classic convective activity for a late summer afternoon in the deep south. Suddenly, the controller and a half-dozen or more airplanes under his control were looking for ways out when there weren't any.

It was obvious this controller had dealt with similar situations before.

"Attention all aircraft, this is New Orleans Approach. Here's what we're going to do. I'm going to drop you all down to my minimum vectoring altitude of one thousand feet and bring you around the west side of the airport for landing to the north. We're switching runways now. I'll start with you, TI three-nineteen. Fly heading two-four-zero and descend to one thousand feet."

"Two-four-zero on the heading and out of three for one, TI three-nineteen," the TI pilot acknowledged.

"Five-five-sierra, maintain two-seven-zero for now and descend to 1,000."

"Out of three for one," I acknowledged.

One by one the controller worked the planes behind me into a new pattern where we all flew a long downwind for landing at the north. He did a good job of keeping us out of the worst of the weather, too.

It gives you second thoughts when an airline pilot flying a big jet doesn't want to be where you are in a single-engine Cessna.

Running the Gauntlet

Date: 7/7/78
A/C Type: Cessna 206
Registration #: N4665C
Route of Flight: Orange County, Texas – Fort Worth

The props on our two turbo-charged Cessna 206s made funny noises in rain. Several of us had experienced this phenomenon and had discussed it during our informal safety gatherings, so this was nothing new to me. It was a strange sensation, kind of like the prop tips were stirring something. The rounded tips on the non-turbocharged 206s and the 210 didn't do it, but the square-tipped blades did.

This doesn't have anything to do with the story I'm about to relate, but I did want to tell you about it, so I figured squeezing this little tidbit of information in here was as good a place as any.

Moreau's pharmacy in Vinton, Louisiana, had a fickle computer system. We visited there often. Many times when I went to Vinton, I landed at the local cropduster strip. You just had to buzz it sometimes to get the cows to move. On this particular occasion, however, I had landed in Orange County, Texas, since there was the possiblilty of IFR weather in the evening and I knew I wouldn't be leaving until after dark.

For the flight home I filed for 8,000, primarily because of head winds and also because the oxygen tank hadn't been serviced. Eight thousand is about as high as I fly at night unless I'm in a pressurized airplane or have supplemental oxygen available.

There was a dead spot in the Lufkin area where there was no radar coverage unless you were above eleven thousand feet. It was a cone in which both Fort Worth Center radar and Houston Center radar both passed right over you.

Just about the time Center reported radar coverage had been lost, I began to see some lightning ahead, some of it to my right front and some of it to my left front. There didn't appear to be any straight ahead. Still I was becoming a little concerned.

Several aircraft on the frequency were asking for weather advisories and were requesting deviations around buildups. I had no storm avoidance equipment, except for my eyeballs. Since I wasn't in radar contact, I couldn't expect much help from Center either.

I flew on, though I was nervous about what might be ahead. I knew from experience that it would probably be at least a hundred miles before the controller would be able to pick me up on radar again. Even though he couldn't see me, I asked about weather in the Lufkin area. There wasn't much help the controller could give me, but he did advise me of signficant weather up around Tyler and Longview.

That wasn't really a concern as I wouldn't be going that far east. But I was also seeing lightning off to the west. From time to time, I lost sight of the ground, an indication that I was flying in and out of, or at least over clouds. Then it grew completely dark as I flew into solid instrument meteorological conditions (IMC), yet the air was smooth.

You're never sure what to do in those circumstances. There was no actual indication that I should turn around. There hadn't been any lightning in my immediate vicinity. It had all been miles ahead. The fact that I was IMC was no big deal, as I flew in the clouds often. I just didn't like the fact that there was no weather information available from a radar controller.

Suddenly, I flew into rain. But it didn't sound like rain. It sounded like hail! Heavy rain on an aluminum airplane can sure make a lot of noise. It was time to turn around, or was it? There was no turbulence, no lightning, just this incredibly hard and loud rain.

I tried the controller again. Still no help. All he could tell me was that he couldn't see any heavy cells in my area and that he should be able to pick me up in another twenty minutes or so.

I flew on, thinking that at any minute I might decide to turn around. The problem was, I didn't know at this point how far the rain extended in front of me. It could be just a local shower that I would be out of in a minute or so.

As it turned out, that was the case. I flew out of the shower and could see the ground again. I could also see lightning ahead and to my right at least a hundred miles away, and some more to my left front, at least a hundred miles away.

"Cessna six-five-charlie, Fort Worth Center, radar contact six-zero miles northeast of Lufkin."

"Roger," I acknowledged. Good to have big brother watching again.

As I reflect back on this story, I think of how far we have come in aviation technology. The early instrument pilots flew in storms and ice with no detection equipment except their eyeballs. During this era in the seventies and eighties, if you flew a sophisticated twin you probably had radar and deicing equipment, but in most singles you were out of luck.

Then came the Stormscope, which changed everything. Now a pilot without radar could "see" storms, sometimes actually see them better than with radar.

Today, the planes I fly not only have stormscopes and radar, but we have Nextrad radar available in the cockpit. It gives you the same view you see on the weather channel. We've come a long way!

Ice is Where You Find It

Date: 11/15/78
A/C Type: Cessna 210
Registration #: N2255S
Route of Flight: Fort Worth – New Orleans – Fort Worth

Fourteen pilots and eight airplanes and I was their daddy. There was work to do in New Orleans and nobody wanted to fly.

"What's the matter?" I asked them.

"Ice," they replied.

I called Flight Service, giving them the 402's N number and type for the weather briefing. No problem, I was told. A little ice in the clouds around Dallas-Fort Worth, but tops were at 4,000 feet. The temperature at Fort Worth was below freezing, but the temperature at New Orleans would be in the 70's by the time I got there.

I gave the guys a hard time, then called Flight Service back and filed an IFR flight plan for 2255S, the Cessna 210. There was no way the boss was going to let me take the 402 on a mere service trip, ice or no ice.

I asked ground control before takeoff to let departure know that I wanted an immediate climb to 4,000. Normally they would hold me at 3,000 until turning me eastbound. After takeoff, the 4,000 foot request wasn't approved, so I trudged along in the soup on runway heading level at three and begin to see a little ice on the leading edges of the wings and on the windshield. It was light rime, but it was building. *Come on*, I pleaded under my breath, *I need higher. Now!* After what seemed like eternity, but was probably only a couple of minutes, I was given a turn towards DFW VOR and a climb to 4,000 feet. At four I was just in the cloud tops. I cheated up a hundred feet and found bright sunshine. It was a dreary winter day below, but pure beauty above the clouds. The ice melted away in 60 seconds.

Departure asked for a PIREP. I told them what I had seen, which was just about what had been forecast. East of Dallas they gave me a climb to 9,000 and a switch to Fort Worth Center all in the same breath. Except for me, the radios were silent in the low altitude sectors.

I flew to New Orleans, did my work, then checked the weather for the return flight. Icing was still in the forecast, but a single PIREP had reported only light rime below 4,000. I realized that it had been me that had made that report hours earlier. I asked the Flight Service weather briefer about it and he told me I had been the only single-engine airplane to fly out of the DFW area all day. Amazing. It was such a beautiful day for flying.

The return flight was uneventful. I flew above the clouds, again in bright sunshine all the way to the outer marker at Meacham. Five-five-sierra slipped down the glideslope picking up hardly any frost at all and landed at a nearly deserted airport. It didn't seem like the 26th busiest airport in the U.S. on this winter day. I made a mental note to write Richard Collins at *Flying* Magazine and thank him for encouraging pilots to go look at the weather instead of relying totally on the forecasts.

The Grand Canyon and The Rockies

Date: 2/20/79
A/C Type: Cessna 402B
Registration #: N98764
Route of Flight: Palm Springs – St. Louis

Palm Springs, California. That name has a ring of exclusivity to it, now doesn't it? Being a pilot involves a bit of exclusivity, too, and so it was that as a pilot I made a trip from Fort Worth to Palm Springs in the Cessna 402 with my friend and fellow V. C. Brown employee, Bill Collins. It was just the two of us. Bill's mission was to make a presentation to a group of Cadillac dealers and my mission, of course, was to get him there and back. Well, not exactly back, because our next stop was in St. Louis and getting there is where the fun began.

The meeting at Palm Springs was at night. The next morning we left early, though we didn't need to be in St. Louis until the following day. With a clear sky and time on our hands, what a great day for sightseeing. We flew straight to Lake Meade and Hoover Dam and from there east along the Grand Canyon at rim level. You can't do that these days without a special permit, but in 1979 there were no restrictions. We flew the full length of the Canyon, turning where it turned to follow the Colorado River until we arrived at the Grand Canyon Airport on the south rim near the east end of the Canyon.

There was snow on the ground when we landed at the Grand Canyon Airport, a rare occurrence. We refueled there and proceeded east across the Rockies to another fuel stop in Wichita.

Crossing the Rockies in the Cessna 402 that day proved to be an interesting experience. The minimum altitude for obstruction clearance along part of our route of flight was 14,000 feet, so we climbed up to 15,000, the proper altitude for eastbound flight, and donned oxygen masks. At 15,000 feet we were flying in and out of cumulus clouds that were holding a good bit of moisture. The left engine started running rough, so I opened the alternate air source in case it was ingesting ice and tried different settings with the mixture, throttle and rpm, but it just wasn't happy.

While I was tweaking the left engine, the right alternator decided to quit. Alternators are like that. They work fine as long as you have a backup, but when it comes time that they might have to carry the load all by themselves, they might very well decide to take a vacation.

There are specific procedures for dealing with an alternator failure in a Cessna 402 and I sort of knew them, but just to make sure, I consulted the emergency procedures checklist in the Pilots Operating Handbook. I also began calling Denver Center and requesting a lower altitude. They let me go down to 14,000, but that was as low as I could go and safely get over the mountains.

Most of the time I couldn't see the mountains, because at 14,000 feet we were in and out of clouds—more in than out. When we were not in the clouds, I didn't like what I saw. We were crossing very jagged mountains with snow-covered tops and they didn't look all that far below us.

Did I mention there was ice in the clouds? Not a lot, but there was some ice starting to accumulate and I was thinking, *I've got one engine about to pack up, and if it does, its alternator will go with it and the alternator on the other engine is already gone and is not responding to all of the recycling and circuit-breaker pulling and electrical load reducing I'm doing and now I've got to cycle the boots and turn on the de-icing equipment? Right.*

I'm begging Center, "Come on, Center, there's got to be someplace we can go where we won't hit a mountain if this engine quits totally and we need to know where it is while I can still talk to you and before I lose electrical power," (though in much more macho language) and Center is telling me, "Just a few more miles and I can get you down." I've got the Cessna 402 Pilots Operating Handbook (POH) in my lap, flipping back and forth in the emergency procedures section trying to find something that will help and Bill Collins is sitting over there in the right seat believing I'm going to get him to St. Louis for his sales demo the next day. This was one of those times they pay professional pilots for, the times when all of your training is going to pay off. At least, you hope it is.

Finally, Center told us we could descend to 11,000 and I did and the left engine started running smoothly and just for grins I turned on the right alternator and it picked up its load and stayed on. Amazing. Wichita, here we come!

Skimming Along

Date: 6/12/79
A/C Type: Cessna 210
Registration #: N2255S
Route of Flight: Ruston – Fort Worth

Two-two-five-five-sierra, depart Scurry heading two-eight-zero, maintain four thousand."

"Five-five-sierra, is out of eight for four, departing Scurry heading two-eight-zero." I acknowledged DFW Approach's instructions, released the altitude hold and beeped the Centurion's nose down with the electronic trim. The deviation from the Scurry 6 arrival was expected when Meacham Airport was your destination.

Waves of light from the setting sun splashed across the cloud tops beneath me, tinting them with traces of red and gold. Home was less than thirty minutes away and the air was smooth and cool. The groundspeed readout indicated no headwind, and my new altitude was going to put me right in the tops of the clouds.

"It doesn't get any better than this," I thought, as I leveled off at 4,000 feet. The groundspeed indicated 160 knots, but it felt like I was flying at jet speeds as I skimmed the cloud tops. The sense of speed was awesome! The clouds were solid and thick, with billowing tops that resembled roll upon roll of tufted cotton and stretched as far as I could see. It was a speed demon's dream.

I could have just flown straight and level. I could have maintained heading and altitude perfectly, letting the autopilot do the work, but what fun was that? I turned the autopilot off, leaned forward in my seat and envisioned a path through the twisting maze. As long as I kept the VOR needle centered and the altitude within 300 feet of my assigned altitude, the controller would never know or care what I was up to.

Picking a towering tuft of white cloud, I headed for it. At the last moment I twisted the control wheel to the left, embedded my wingtip in the cloud and skirted its perimeter. Coming around on the opposite side, I dove into a slight valley of cloud, then climbed up the far side. It was "nap-of-the-clouds"—contour flying at its best.

Over the misty hills and into the foggy valleys the 210 raced, nip and tuck with the clouds—a roller coaster ride, free from rails and trestles. Twisting around the peaks, burying a wing here, punching through a cloud there, I was having the time of my life. In the radar room at DFW my blip on the scope tracked steadily toward Fort Worth. There was no way they could know the fun I was having.

The course change came all too soon. "Five-five sierra, fly heading three-three-zero, maintain three thousand."

"Five-five-sierra is out of four for three, heading three-three-zero."

Into the clouds I went, deeper and deeper, until the white turned to gray and I was solid on the gauges. It was time to put "George" to work again. Heading bug to 330, autopilot on, altitude hold on at 3,000, it was serious flying now. Approach charts out, the ILS and Outer Marker tuned in, ground speed readout no longer accurate as I tuned number two VOR to DFW and locked the DME to it. There was darkness below, as the sun had followed me into the clouds in a spectacular sunset. The descent clearance to two thousand came and the darkness began to come alive with the twinkle of lights as I flew out of the bottom of the clouds to the sight of hundreds of thousands of Metroplex drivers making their way home. Think I would trade places with them? Not on your life!

Pressurized

Date: 8/20/79
A/C Type: Cessna 421C
Registration #: N234CC
Route of Flight: Fort Worth – Montrose, CO – Denver – Fort Worth

Becki called me at the hangar and told me to go up to Red's Air-craft and catch the boss. She said he had just bought a Cessna 421 and was about to take off for Colorado in it. "Go with him," she said. Becki was the other multi-engine pilot and the boss's friend. She was con-cerned that V. C. Brown, who was a private pilot with mostly fair weather experience, was about to take off on an extended cross country in a new pressurized airplane that he had never flown before, and that he was going to do it without her supervision. She couldn't go, but perhaps I could.

I went up to Red's and found V. C. and one of our salesmen in a brand new 421. I didn't even know he had bought the new plane. The company salesman who was in the copilot's seat was not a pilot. I opened the door and stuck my head inside, just as V.C. was about to start en-gines. "Can I go?" I asked.

"I guess so, but you can't fly," V. C. grumbled. I had no idea where we were going and had no clothes, not even a toothbrush, but I climbed on board and shut the door.

"Where are we going?" I asked as V. C. taxied to the runway.

"Colorado," he told me.

Forty-five minutes later, we were flying along the airway toward Amarillo, when some ominous-looking clouds appeared ahead. V. C. called me to the cockpit. "You want to fly us through this stuff?" he asked.

"Sure," I replied, eager to get in the left front seat of the new pressurized twin. V. C. put the plane on autopilot and got out of the pilot's seat. I slipped in, buckled up and began to look around. "What's our clearance?" I asked.

"I didn't file," he replied.

We were about to get into some pretty dark clouds, so I dug around for some IFR charts and looked at the routing. "What's our destination?" I wanted to know.

"Montrose, Colorado," the boss told me.

I got us a clearance, then began to study the airplane. It had all kinds of new goodies that were new to me. To start with, there was an HSI, a Horizontal Situation Indicator. I had read about them, but we didn't have one in any of our other airplanes. There was also a new radio called RNAV for "area navigation" that would allow you to off-set navigation points from the airways, thus creating a direct route.

I flew that airplane around in Colorado and Texas for the rest of the day. V. C. never asked for the left seat back, because the weather was IFR most of the time. We went to Montrose, then to Denver. Flying into Denver, IFR at 19,000 feet (Flight Level 190) I was given a holding clearance. Because of the HSI, holding would be a piece of cake. I asked for five mile legs; the controller gave me ten mile legs.

The pressurization was pretty straightforward and I figured out how to use it based on the information in the airplane manual. I played around with the RNAV and by the time we left Montrose, was already comfortable filing and flying direct.

Denver was IFR when we went in there, so I flew a coupled ILS approach with the autopilot, which worked out very nicely. Leaving Denver, I again filed for Flight Level 190, just because I could, and because that altitude had very favorable winds and got us above most of the weather. I was in the big time now.

In the pilot's lounge in Denver, I had run into an old friend, Jim Gibson. Jim had left V. C. Brown a year or so earlier and was flying a King Air 200 for a Fort Worth Oil Company. Since he was on his way back to Fort Worth, too, we agreed to monitor 122.9 on our second radio for some conversation en route.

I had just leveled off at FL 190 when Jim called me. He was a few miles ahead of me at FL 210. "Dave," he said. "There's a little puffy cloud ahead of you a few miles sitting all by itself. You'll want to avoid that cloud."

I saw it a few minutes later, right on my course and at my altitude. *Avoid it*, I smirked to myself. Such a little cloud, so harmless. Punching through it would be fun. I pointed the nose of the airplane right at it. We would punch through it in nothing flat.

Fifteen seconds later, the 421 was almost upside down. I should have listened. How could such a small cloud pack such a wallop? "Sorry," I told my passengers.

The Flight to Remember

Date: 3/11/80
A/C Type: Cessna 421C
Registration #: N234CC
Route of Flight: Greenville – Little Rock

The day started all right with an assignment to deliver two computer systems—one in Louisiana, the other in Mississippi. A note in the airplane said the autopilot was inoperative. No problem. There would be five or six hours flying at the most. It wouldn't hurt me to do a little hand flying every now and then.

We loaded up and departed Meacham a little after 9:00 a.m. A couple of hours later we were in Lake Charles where I dropped off one computer system and the technician that was going to install it. Then it was off to Greenville, Mississippi, for the next stop. Thunderstorms were starting to build up, typical for March in that part of the country, but there was plenty of room to go around them and the airplane's radar was working fine.

By the time we had landed in Greenville, a line of thunderstorms had formed just to the west of the airport and was moving through the area. It seemed prudent to wait them out, so I decided to stay and help the installer get the computer system up and running. It would take a few hours, but that was okay, as it was starting to rain and there was no shortage of thunder and lightning.

It was around 11:00 p.m. when the installation was finished and the pharmacist took me back to the airport. The installer was going to stay for a few days and train the pharmacy staff on how to use the new computer.

It wasn't raining when we got back to the airport, but the sky was still cloudy and the ramp was wet. Off in the distance I could see an occasional flash of lightning. The pharmacist drove off and I put the tools and excess gear from the installation in the plane, then went off to find a phone to call Flight Service for a weather briefing.

I couldn't find a phone, which was ridiculous because airports always have a pay phone outside where you can get to it in case you fly in during the middle of the night. But if there was one at this airport, I couldn't find it, and the FBO was all closed up for the night. The airport appeared deserted. I decided to try to reach Flight Service via the radio in the plane.

"Greenwood Radio, this is Twin Cessna two-three-four-charlie-charlie, listening Greenville." The latter part of the message indicated to the Flight Service Station at nearby Greenwood airport that I was at Greenville and the best way to reach me would be via their remote transmitter located on the ground at Greenville. They didn't answer. I tried again, and still no response.

Switching frequencies, I tried the FAA Air Traffic Control Center that was responsible for the area. "Memphis Center, Twin Cessna two-three-four-charlie-charlie, on the ground at Greenville."

"Two-three-four-charlie-charlie, Memphis Center, go ahead," came the instant reply.

"I've been trying to reach Flight Service with no luck," I told the controller. "I'm a Cessna 421 wanting to go IFR to Fort Worth, and wondering about the current weather situation."

"Cessna two-three-four-charlie-charlie, I'm painting a solid area of thunderstorm activity as far west as I can see on my radar. There's a pretty good cell just west of you that will be moving over the airport within a few minutes."

"Okay, I'll try again in the morning," I replied. "Good night."

"Good night, four-charlie-charlie."

When I opened the door to the airplane, rain was beginning to splatter upon the ramp. It was then that I saw the heavy lightning associated with the storm the controller had told me about. I decided to try again to find a phone to call a motel. This was in the days before the cell phones we now take for granted.

No sooner had I closed up the airplane and started across the ramp than it began a heavy downpour. I ran for the shelter of the closest building, but got soaked. As I ran along the side of a hangar, trying to get some shelter from the rain from the slight overhang of its roof, I saw the lights of a pickup truck coming through a gate and onto the ramp. I stepped away from the building and flagged the driver down. By now I was soaking wet.

It was a security guard and he drove me over to the fire station so that I could use the phone. I spent the next few minutes calling one motel or hotel after another looking for a room. There was some kind of convention in town and the rooms were all booked, every one of them.

No problem, I figured. The Cessna 421 has a roomy cabin and comfortable seats. I decided to just sleep in the plane. Storms as violent as the ones that had been raging all afternoon and were now passing through normally move through very quickly that time of year, leaving clear and windy skies behind. If I could manage to sleep a couple of hours, the weather would probably be fine and I could make it home, which was only two hours away in the 421.

There were some snacks in the plane, so I wouldn't go hungry. The seats were comfortable enough, but what I hadn't counted on was the cold. It was March and I was soaking wet. I was soon very uncomfortable, unable to shake the cold. I began to think about my father's house in Jackson, or the Holiday Inn in Oxford that would surely have a room, or one of the motels right at the airport in Memphis. I called Memphis Center again.

"Memphis Center, Twin Cessna two-three-four-charlie-charlie again. What does it look like down towards Jackson?"

"Twin Cessna two-three-four-charlie-charlie, Memphis Center, I show a pretty solid area of medium-to-high intensity cells between Greenville and Jackson, extending into the Jackson area."

"What about Memphis?" I asked.

"We have a thunderstorm at the airport now. There are still quite a few cells between your location and Memphis."

That ruled out Oxford, too, since it was between Greenville and Memphis. Besides, the only instrument approach into Oxford was to a non-directional beacon that belonged to Republic Airlines. You had to be practically over the field to pick it up.

I started the left engine and fired up the cabin heater to warm up the cabin. If I could get dried out and the cabin got warm enough, I might get some sleep.

The cabin warmed up all right, but as soon as I turned the heater off it got cold again. I just couldn't get comfortable enough to go to sleep. Every time there was a break in the rain or it seemed the lightning had stopped, I cranked the left engine to supply electrical power, turned on the radar and scanned the western sky. Each time there were cells up there—big nasty ones.

Finally, I dozed off. When I woke again around 2:00 a.m. it wasn't raining. I watched and listened for lightning or thunder and there didn't seem to be any. I called Memphis Center again.

"Memphis Center, Twin Cessna two-three-four-charlie-charlie on the ground at Greenville. I'd like to check the thunderstorm situation going out to Fort Worth."

"Two-three-four-charlie-charlie, Memphis Center. Most of the weather has moved on through and it looks clear out to the west as far as I can see."

I scrambled for my charts and a pad and pencil. "Two-three-four-charlie-charlie would like to file an IFR flight plan with you then."

"Go ahead with your flight plan," the controller replied.

Normally, you don't file a flight plan with Center, but it was the middle of the night and the guy wasn't too busy, so he was accommodating.

I knew better than to give in to "get-home-itus" the pilot-killer disease. I knew better than to let wanting to get home to a comfortable bed push me into flying when I really shouldn't. After all, hadn't I waited on the ground for several hours because the conditions weren't favorable for flying? Now everything was fine; the controller had just told me so. He said it was "clear" out west. That fit in with what I expected. It usually cleared behind a fast moving cold front. I was cold, tired, and miserable in my wet, dirty clothes, and home was only two hours away. I filed the flight plan.

But I had missed something. The controller had told me it looked clear to him out west. But I had only asked about thunderstorms. That was my mistake. I should have requested a complete weather picture for the flight home. The "clear" that I heard was what I expected to hear, what I wanted to hear, but it wasn't what he actually meant, as I was to discover later. The controller was talking about thunderstorm activity as he saw it on his radarscope. The radarscope was "clear" of thunderstorm activity. I thought he meant the stars were shining out west, but I hadn't asked the right question.

Although it was still cloudy at the airport, I was convinced that it was just a local condition and that I would have VFR conditions as I flew toward Fort Worth. I knew I was too tired to fly two hours on instruments, then fly an instrument approach to minimums, but VFR? I could handle VFR.

I filed RNAV direct to Blue Ridge, Blue Ridge 6 arrival to Fort Worth Meacham, 10,000 feet, two hours en route, four hours fuel on board. Memphis Center came back with a "cleared as filed" and I taxied out to the runway. The Twin Cessna's heater was already starting to warm my cold-soaked bones.

The first sign of trouble was when I went into the clouds right after takeoff. I hadn't even gotten the landing gear retracted when I found myself in solid instrument conditions. I turned off the wingtip strobes to avoid vertigo and concentrated on keeping the wings level as I climbed. I was having trouble with my instrument scan and the airplane was ahead of me—way ahead of me. Not good. The radarscope was alive with images, but many of the black spots I thought were thunderstorms were actually returns from the ground as I allowed the airplane's wings to rock from one side to the other.

My major enemy was not thunderstorms; it was fatigue. How could I have been so tired and not realized it? Turning on the autopilot to square everything away and give myself a chance to catch up with the airplane wasn't an option because the autopilot didn't work.

I concentrated on flying the attitude indicator, forgetting all of the other instruments for the moment except the altimeter. I had to climb so I concentrated on trying to hold the nose slightly up and the wings level, figuring the rest of the stuff would take care of itself until I got some distance between me and the ground.

I began shaking uncontrollably. This was a type of fear I had never experienced before, not even in Vietnam. I thought I was going to die. Every pilot who has flown for any length of time will tell you they've had times when imaginary headlines start going through their head. You know the kind: "Texas Pilot Dies in Fiery Crash" and so on. Now imaginary headlines raced through my mind as I held panic at bay with pure willpower. I forced myself to put this kind of thinking out of my head and remembered my old instructor Jim Miles' words, "Never give up when you still have some control left. Fly the aircraft! Fly it even if you have to fly it into a crash." I had wanted to fly this airplane. I had coveted the left seat of this sophisticated twin and now it was going to be my coffin if I couldn't master it.

I began talking to myself. "Come on, David, level your wings, positive rate of climb. Fly the airplane. You can do it." Slowly, I began to mentally catch up with the plane, but I was too tired to stay with it for long. My first thought was to come back around and land, but to do that I would have to fly an instrument approach to minimums. Greenville was at that time a no radar environment, which meant that I would have to fly a full approach with a procedure turn and with no guidance from the controller. I just wasn't up to it, and I knew that I wasn't.

I did the only other thing I could do, which was turn on course, climb to my assigned altitude of 10,000 feet and concentrate on just flying the attitude indicator until I got into VFR conditions. I couldn't calm myself down, couldn't relax. My hands were sweaty and I was trembling. I was shaking, as if I were freezing cold, though the cabin was quite warm. I was in way over my head, but I couldn't give up.

Things continued to go wrong. When I reached 10,000 feet, I was still in the soup. The images on the radar were no longer images of the ground. There were real cells out there, a lot of them. Maybe they weren't intense enough to show up on the Memphis Center's Air Traffic Control radar, but they were more than I wanted to fly through that night.

As I leveled off and tried to maintain something resembling straight and level flight, I decided it was time to get some real weather information.

"Memphis Center, Twin Cessna two-three-four-charlie-charlie, can you get me some weather information for my destination and along my route of flight?"

"Twin Cessna four-charlie-charlie, stand by."

When the controller came back on the air with the weather report, my spirit hit bottom and my knees began shaking so violently, I couldn't control them. "Fort Worth Meacham is reporting ceiling four hundred feet, visibility two miles, rain and fog. Dallas Love Field, ceiling 600 feet, visibility two miles, rain and fog. Dallas-Fort Worth Regional Airport, ceiling 400 feet, visibility a mile and a half, fog. Texarkana, sky obscured, visibility one-half mile, thunderstorms.

That's when it dawned on me that the controller's meaning of the word "clear" and my meaning of the word "clear" were not one and the same. As I was digesting the weather information and silently praying to God to get me on the ground in one piece, the controller gave me a frequency change. I decided to confess my predicament to the new controller.

"Memphis Center, Twin Cessna two-three-four-charlie-charlie is with you at one-zero thousand."

"Twin Cessna two-three-four-charlie-charlie, Memphis Center, roger."

"Memphis Center," I began, "Two-three-four-charlie-charlie, my destination still has low ceilings and visibility and I'm just too tired to fly IFR for two more hours. What's the weather like up around Memphis now?" Memphis was about a 100 miles north of my location.

"We've got 800 foot ceiling and two miles visibility at the airport, but there is still some pretty solid weather between your location and Memphis."

"What about Jackson or Monroe?" I asked. They were both south of where I was.

"Well, let's see . . . Jackson is showing a thunderstorm at the airport on the last hour with numerous buildups in all quadrants. I don't have anything current on Monroe, but it still looks like some pretty heavy precip down that way." What was this controller seeing that the other one hadn't seen, I wondered. Oh, yeah. He had said, "as far as he could see." Apparently, that hadn't been very far.

Great! I was too tired to avoid the cells I was seeing on my own radar. In spite of the intensity of some of the echoes, the turbulence wasn't destructive. It just made it difficult for me to keep the plane on a steady course. I pulled my seat belt down tighter, slowed to the maneuvering speed of 151 knots (as close as I could maintain it), and concentrated on keeping the wings level. The altitude was more than I could handle. There were too many ups and downs, but there wasn't anybody else out there for me to run into, so the controller didn't bother me about it.

"What about Little Rock?" I asked.

"Little Rock looks like it might be your best bet," the controller replied. "They have 1300 foot overcast with two miles of visibility. There is only one cell near there big enough to show up on my radar and it is presently fifteen miles southwest of the airport."

I didn't hesitate. "How about a clearance direct to Little Rock and a lower altitude?" I asked.

"Twin Cessna two-three-four-charlie-charlie, you're cleared from your present position, direct Little Rock. Descend and maintain six thousand," the controller responded.

"Two-three-four-charlie-charlie is out of one-zero thousand for six thousand, direct Little Rock." I repeated my clearance and breathed a sigh of relief. I tuned in the Little Rock VOR and saw the needle quiver slightly, then settle down, giving me the magnetic course to the airport. The DME (Distance Measuring Equipment) indicated the airport was seventy-four miles away.

I might just make it. At six thousand feet, the ride wasn't too bad. Now all I had to do was psyche myself up for the approach.

A few minutes later when I was handed off to Little Rock Approach I was flying between cloud layers. The layer beneath me was pretty thin and I started seeing lights on the ground. I saw the glow of Little Rock when I was still ten to fifteen miles away. Approach Control began vectoring me for an ILS approach and pointed me right at the only storm cell that was visible on my radar. I asked for a heading direct to the outer marker and as I got in closer, I could see the airport through breaks in the clouds. I requested and was cleared for a contact approach.

A few minutes later I was on the ground. I taxied up to Central Flying Service, got out of the airplane on shaky legs, kneeled down on the wet pavement and kissed it. Within a few minutes I was in a warm, cozy hotel room, where I slept until mid-morning.

Chapter 6 – Another Day, Another Company

V. C. Brown, Inc. went out of business in May 1980. I didn't know it, but the company had never made a payment on the Cessna 421. Mr. Brown had me taxi the airplane up to Red's and leave it with the keys in it. Repossession of the other airplanes soon followed.

I joined forces with some guys I had recently met who had a pharmacy software package with a lot of promise. We started a new company, TBL, Inc. TBL stood for The Business Language, a modern day programming language that we knew would take the world by storm. Dennis Smeltzer, one of the pilots that I had hired at V.C. Brown, came on board. We didn't fly right away because we were struggling to build a business. But, as the business began to grow, we started renting a Cherokee Six, then we bought one. A few months later we traded a computer system and some cash for a Lance, which is a retractable version of the Six. Dennis and I were the pilots.

We operated TBL for three years, then sold it to a big pharmaceutical company. We kept the planes for a while and the parent company bought a King Air. I got to fly it a few times before being relegated to a desk job.

The pharmaceutical company didn't do a very good job with TBL. Within a couple of years they were ready to get out of the computer business. They started selling off everything, including the airplanes. The Cherokee Six had originally been bought by one of the principles of TBL and leased to the company. It was turned back over to him. The Lance was sold to another employee for a measly $14,000. I didn't buy it because I was promised an interest in the Cherokee Six for nothing. Big mistake, as the registered owner of the Six reneged on his deal before a year was out. The guy who bought the Lance couldn't afford to insure it, but he sold it for a hefty profit.

There are a few good stories from this era when Dennis and I were flying around the country in the Cherokee Six and Lance trying to keep pharmacy computers running.

2500', No Lights, at Night

Date: 4/24/81
A/C Type: Cherokee Six
Registration #: N2881G
Route of Flight: Wynne, AR – Salem, AR

There are country pilots and there are city pilots. I was a country pilot, Dennis was a city pilot. Country pilots learn to fly out of 2500 foot grass strips, 30-feet wide, that may or may not have runway lights and may or may not line up with the wind. City pilots learn to fly out of airports with multiple concrete runways, 5,000 feet long or longer, 100-150 feet wide and well lighted.

Country pilots do things routinely that city pilots aren't used to doing. So it was that when Dennis Smeltzer and I, together in Wynne, Arkansas, got a call from a customer wanting us to come to Salem, Arkansas, after dark, I said, "Sure."

Dennis said, "What? No way!"

Dennis picked out a lighted, paved airport in Cherokee Village, about fifty miles from Salem and said, "We'll go there."

I asked the customer about the 2500 foot, unlighted strip that was in Salem. "How can we find it?" I wanted to know.

"Approach town from the east and fly west along the main highway until you're directly over the honky-tonk that's just outside of town. Then turn south right over that honky-tonk and start descending. The runway is about two miles south of there. I'll put some lights on the north end of the runway and light up the south end with my car lights," the customer told me. Obviously, he knew a little bit about flying himself.

"Sounds good to me," I said.

"We'll call you from Cherokee Village," Dennis said.

I had to cut Dennis some slack. He hadn't spent seven years in the Army flying helicopters into unimproved, unlighted areas, and he hadn't learned to fly at Oktibbeha Airport in Mississippi, where the runways were grass and where there were no lights. He had learned to fly at Charlie Brown Airport in Atlanta. But it was my leg to fly, so I told Dennis that we would try Salem first and if it didn't look good, we would go to Cherokee Village to land.

It was extremely dark in Arkansas that night, especially up in mountain country. But I did find the highway and I flew along it at fifteen hundred feet and found the honky-tonk. Over the honky-tonk I turned south and started descending. The airport elevation was 700 feet, so we had 800 feet to go. I told Dennis to call my altitude at hundred foot intervals and if we didn't see the airport by the time we were 300 feet above the ground we would abort and go to Cherokee Village.

I put on the first notch of flaps, slowed the airspeed to 90 knots and turned on the landing light. I didn't see anything to mark the airport at first, but then I saw a flashing yellow light that I hoped was marking the near end of the runway. It looked like one of those construction barrier warning lights. Then a pair of headlights came on, pointing our way. Those would be at the far end of the field. The distance between them looked mighty short. If that was the runway, it looked like the top of an aircraft carrier. The chart said it was 2500 feet long, so I had to believe it.

"Five hundred feet," Dennis called. I pulled up the lever to add the last two notches of flaps. I was in the groove with a good sight picture. We would touch down just beyond the first set of lights.

"Three hundred feet, go around," Dennis said. He was hedging. I never heard a four-hundred foot call, but I wasn't going around. Everything was looking good as my landing light illuminated the ground and I saw that there was grass in front of us instead of trees.

When we touched down, I got on the brakes a little too hard and dust flew everywhere, I couldn't tell how much room we had and wanted to make sure we got stopped before running out of runway. The Six skidded to a stop halfway down the runway, the only bad part about the landing. When we stopped, the car lights were still hundreds of feet in front of us. A piece of cake, routine stuff. Dennis thought I was crazy.

Two-Thousand-Nine-Hundred and Ninety-Nine Feet of a Three Thousand Foot Runway

Date: 12/15/81
A/C Type: Cherokee Six
Registration #: N38276
Route of Flight: Decatur, IN – Denton, Texas

Just before Christmas I learned that the company had sold a computer system to a pharmacy in St. Marys, Ohio. Since my wife's sister and her family lived near there, I volunteered to install the system with the understanding that my family would go along to visit relatives. It was a good deal for the company, because it meant they wouldn't have to pay for me to stay in a motel.

Our family now consisted of Jamie, our oldest; Nathan, who was born in Oxford; and Joyce was expecting our third child. We flew up in the Cherokee Six on December 9th and landed at a grass strip right on the Indiana-Ohio border, just outside of Decatur, Indiana. During the time we were there three or four inches of snow fell. When it was time to return to our home in Texas, the snow was still on the ground.

Being a southern boy, I wasn't really sure what effect the powdered snow on the ground would have on my takeoff roll. I knew to get all of the snow and frost off the wings of the plane. I did that, then went to talk to some local pilots to find out what to expect on takeoff. The only person I found was a man in the airport office who appeared to be in his eighties. I told him my concern and he said, "Oh, I wouldn't worry too much about it. That airplane has plenty of lift." Since he was obviously a man of great experience and since it was true that the Cherokee Six was an airplane with plenty of lift, I concluded that the takeoff would be safe.

I phoned Flight Service to get my IFR clearance, along with a void time that necessitated us being off the ground in ten minutes. The weather was overcast with a few breaks beginning to open up in the overcast. We were lightly loaded and it appeared we would be able to get on top of the clouds quickly enough that ice wouldn't be an issue. The temperature on the ground was well below freezing.

There was a north wind of around ten knots, so I taxied down to the south end of the runway. The plane seemed to taxi okay. A power line was stretched across the north end of the airport, right at the end of the runway. I didn't like that power line. In fact, it made me very uncomfortable.

I gave the engine plenty of time to warm up and when the oil and cylinder head temperatures were well within the green ranges, I did a normal runup and everything checked out. Using my best soft field technique, I started the takeoff roll. Our clearance void time was approaching, so if I didn't get off quickly I would have to shut down, go inside and call Flight Service on the phone again to get another clearance.

The takeoff roll was really slow and the power line loomed large in my thoughts. Halfway down the runway we hadn't obtained flying speed, so I aborted. It was then that I made the decision to make the takeoff going the other direction. Yes, it would be a downwind takeoff, but there was a big open field at the south end of the runway in case we didn't make it off for some reason. The only negative was a small farm road running perpendicular to the end of the runway. If we ran off the end of the runway before getting airborne, we might do some landing gear damage, but at least we wouldn't be tangling with a power line.

Roaring down the runway at full power, the airspeed seemed to hover around 50 knots. We needed 60 to 65 to break ground and begin accelerating. The point of no return passed with the airspeed just under 60. The Six wouldn't fly yet, but she was gaining momentum.

Suddenly, a tractor appeared on the road in front of us and it appeared we were on a collision course. Just in time, the driver saw us and stopped. The airplane had to fly or it was going to wind up in the field, possibly with a bent prop and who knows what else. I sucked in my breath, talked nice to the Six and she broke ground after using up 2,999 feet of that 3,000 foot runway.

Once airborne, she was in her element, and we climbed normally. We were IFR for a while and the cockpit workload was high, but within an hour we were cruising along on top of the clouds in beautiful flying conditions. I had planned a fuel stop for Cape Girardeau. When the plane touched down on the runway there, the wheels didn't want to turn. It was as if the brakes were locked. We skidded down the runway and I expected a blowout at any moment. Finally, the wheels broke free and I was able to taxi the plane to parking.

After shutting down on the parking ramp, I went to examine the wheels and tires to see what was wrong. Both wheel rims were full of ice that was frozen solid. Since the airplane had wheel pants that covered the wheels this would have been very difficult to see at Decatur when the plane was sitting in approximately four inches of snow. It was no wonder the takeoff roll had been difficult. The ice was acting as brakes. It was as if we had made the entire takeoff roll with the brakes on, so the wheels were sliding in the snow instead of rolling.

That was something I had never encountered before, and had never been told about, to my recollection. It was one more thing learned by experience, rather than by formal training.

Total Electrical Failure

Date: 7/11/83
A/C Type: Piper Lance – PA32R
Registration #: N1978J
Route of Flight: St. Louis – Charlotte Wilgrove

I had to drop a system off in St. Louis, then go to Raleigh-Durham for a user meeting. It was a great opportunity to take Joyce and the boys (we now had three) over to Charlotte to visit her parents. The flight to St. Louis took most of the morning and I had to spend a little time there, so it was after dark when we arrived in the Carolinas.

There was a cloud layer beneath us, starting around Chattanooga. The Charlotte weather was reported as 2,000 feet overcast. We were at 7,000 feet and between layers east of Asheville, when I began to smell something funny, like sulphur. All of the gauges looked okay, including both the voltmeter and ammeter, but I didn't like that smell.

We were given a descent clearance and handed off to Charlotte Approach. I reconfirmed with the Approach controller that the ceiling was still 2,000 feet. So far, I didn't have anything to report to him in the way of an emergency, but the sulphur smell had stayed with us and had to be coming from the airplane.

I asked for vectors to Wilgrove Airport, which is where Joyce's parents were expecting us. When we broke out of the clouds, it was very hazy beneath with limited visibility.

Wilgrove had never been easy for me to find in the dark. It didn't have a rotating beacon and you had to be practically right over the airport to pick out the runway lights. I strained to locate the airport in the darkness, anxious to be on the ground and find the source of the strange odor.

When I finally did see the airport, I was practically in the traffic pattern. I entered downwind, and lowered the landing gear. Operation of the gear switch resulted in a total electrical failure. Not only did the panel become dark, but I couldn't tell if the landing gear was down or not. I had no landing light and no radios. Visibility was too poor to be wandering around in the sky with no lights, so I continued the approach. As we turned final, I pulled the emergency gear extension handle and hoped for the best.

Fortunately, I had practiced night landings without lights many times and was comfortable using sound and the airplane's attitude to determine the proper approach speed. I was reasonably confident the gear was down because it free falls into position when the emergency release handle is pulled. I had heard a solid thunk, when originally putting down the gear, but I didn't have the three green indicator lights to assure me the gear was down and locked.

As we touched down, I felt gingerly for the ground and was very relieved when I felt wheels make contact with the runway instead of the propeller and the aircraft's belly. Finding a parking spot without a landing or taxi light was challenging, but with the help of a flashlight, I managed to find a tiedown spot.

The next morning, I opened up the battery compartment, which is in the small luggage compartment between the instrument panel and the engine, just ahead of the firewall. There I found the source of both the sulphur smell and the electrical failure. The battery had overheated, swelling almost to the point of exploding. The mechanic that helped me install a new battery told me that it was because the water level in the battery had been too low. The battery in a Lance is a little hard to get to, so checking its water levels is something normally done on maintenance inspections. As a pilot, I had never checked it. Consequently, I could have ruined the whole airplane, not to mention my family. Check your battery periodically, you who fly your own airplanes.

A Closer Look at Rainbows

Date: 7/6/86
A/C Type: Cherokee Six
Registration #: 38276
Route of Flight: Goode - Local

Cruising above the Dallas skyline in search of rainbows that had moved on or no longer existed, we found the clouds had all moved off to the east. Not a rainbow was in sight. At our backs the sun was dipping into the haze, changing colors as if the gray layer of sky were a translucent bucket filled with red paint. Phillip's nose was pressed against the plexiglass window, his eyes taking in the sights below. That boy doesn't miss much.

Soon it would be dark. Thirty minutes earlier we were in our front yard watching rainbows form along the trailing edge of a line of thundershowers. "Daddy, when you fly through rainbows, do you see the colors?" It was the kind of question that four-year-old minds ponder. For me, it was an excuse to fly. Heck, anything was an excuse to fly. I could have given a quick answer to Phillip's question (you don't fly through rainbows; they're as illusive from the air as they are from the ground), but I chose to show him. "Let's go find out," I said to my youngest son.

The airport was three miles from the house. Three years earlier, when we had moved our plane there, it had been a grass strip, home to twenty to twenty-five planes. At the time of this flight, Goode Airport had a paved runway and multiple rows of hangars filled with aircraft of all ages, types and colors. Our Cherokee Six was parked in a hangar at the bottom of the hill, near the south end of the runway. As we drove around the perimeter road, Phillip pointed to, asked about, commented on, and admired the planes we were passing. We lived in an airline pilot community. There were Bonanzas, assorted Cessnas, Cherokees, and several antiques and classics. I loved the fact that one of my sons shared my love for these machines. Phillip's two older brothers got excited about Porches and Ferraris. I was happiest around old machines that were made in America when Americans made fine machines.

The Six waited to serve us faithfully, as it had for business and pleasure for half a decade. Ten years old, the six-passenger plane had cost just slightly more than our custom van. It wasn't as roomy and cost a lot more to operate, but when you were going to visit the grand-parents in the Carolinas, five hours sure beat two hard days.

Putting the Six into the air that evening was like releasing a fish into the water. It was at home. We flew over the Dallas-Fort Worth Metroplex in search of the answer to a little boy's question, and I real-ized that most of the three million people living and working below were oblivious to our presence. Too bad for them. Phillip and I were enjoying ourselves immensely. Statistics indicate that less than 15,000 of the people living in the communities below were pilots. Perhaps the number is a little higher in this community, headquarters for a major airline and a major hub for several others. Still, a very small percentage of the overall population holds a pilot's license.

A half a century earlier, Walter Beech, Clyde Cessna, William Piper, and others whose names are no longer associated with modern aviation, believed there would come a day when airplane ownership would be as common as the family car. At various times in our country's history, it seemed their dream would become a reality. Yet, even now the Brother-hood of the Sky remains exclusive. For the most part Americans are indifferent or apathetic concerning what could be one of their greatest freedoms. Mention flying to them, and they automatically think of the airlines.

I wouldn't trade anything for the look on Phillip's face, as he turned to me with wonder in his eyes. We banked to follow the loop around the south edge of the city and I pointed out Dallas Love Field and Navy

Dallas to Phillip. I dimmed the instrument panel lights, noting that the familiar gauges on the instrument panel were indicating exactly what they should to reflect the inner workings of a machine that was performing exactly as it should a task that I have repeatedly taken for granted.

By the time we returned to the airport, an hour had ticked by on the Hobbs meter and the runway lights were on. We landed uphill, the tires kissing the pavement with the familiar "chirp, chirp" that signaled a textbook arrival. Phillip smiled and gave me a thumb's up. He knew a good landing. He also knew what a bad landing was like.

Later, with the airplane put to bed and Phillip pestering his mom for something to eat, I sat down at the computer to work on the _Highways in the Sky_ book with new resolve. There was something about getting this particular book written that was akin to the determination with which I pursued a job in corporate aviation years ago. At a time when the aviation periodicals declared that there were "10,000 qualified pilots for every pilot's job available," I continued to look at every airplane that flew over, telling myself, "There's somebody in that left seat. It could just as well be me." I knew that stories about flying were worth telling, and I believed the storyteller could just as well be me.

This flight occurred and the story was originally written when Phillip was four years old. He is now twenty and just a few hours short of obtaining his own pilot's license. The fact that it has taken so long to get the book written is another story altogether.

Chapter 7 – Resurrection

Sometimes logbooks tell you more by what is not in them than what is written in the lines that fill their pages. Mine are like that. From late 1970 until the end of 1979, each page in my logbook covered just a few days, indicating a lot of flying. Beginning in 1980 and progressing through the end of 1985, one page in the logbook often covered a period of two or three months. Pursuits other than flying were obviously occupying much of my time. On December 30, 1985, the entries stopped abruptly.

I remember that last flight well. Before Christmas, I had flown to Van Wert, Ohio, to pick up Joyce's sister Jeannie, her husband Dennis, and their children so that they could spend Christmas with us. After Christmas I flew them home.

It was bitter cold when I left Van Wert the second time and the airplane's compass was acting up. Flying towards Indianapolis, the heading on the compass showed that I was flying toward Chicago. Fortunately, the navigation radios were working fine and all agreed with each other, so I knew I was going toward Indianapolis. Still, it gives you an eerie feeling to be flying along with something so basic as a compass trying to steer you wrong. It finally straightened up and began giving indications I could trust, but it took more than an hour.

Once the compass seemed reasonable, I aligned my directional gyro with it and navigated on with reasonable assurance I would make it to Texas. It took eight hours to get home. I stopped for fuel in Springfield, Missouri, and filed an IFR flight plan for the rest of the journey home. It's a good thing, I did, too. The flight ended with an ILS approach to minimums at Denton.

A few weeks after that, the Six was gone. It was supposed to have been a partnership deal, but the plane was in one person's name and he was the nonpilot member of the partnership. He got rid of the Six in early 1986. He didn't even give it the dignity of selling it, but traded it for a Jaguar sports car.

In some ways having the Six gone was a relief. I couldn't really afford to fly it or maintain it, and even the insurance premiums were a struggle, but the downside was that I stopped flying. I never meant to stop flying, it just happened. It had been years since I had rented an airplane and I didn't have any contacts in the aviation business any more. My flight instructor certificate expired, then my medical and I just let them. Life had enough challenges just trying to make a living and raise a family.

Jamie grew up and got married, and Nathan soon followed. Only Phillip was still at home and he was eighteen and out of high school. One day in the Fall of 2000, Phillip told me he wanted to learn to fly, perhaps even become an airline pilot. This announcement caught me totally off guard, but needless to say, I was delighted. We went looking for a flight school.

The airport where we had kept the Six had closed and houses were being built where the runway and hangars had been previously. There was an airport not too far from where we now lived that I had used from time to time when I was an active pilot. In those days it was called Aero Valley, but it was now known as Northwest Regional Airport. Phillip and I drove out to Northwest Regional and found that there were two flight schools on the airport. We picked one and signed Phillip up for a lesson.

I decided that I wanted to get current, too, but had apprehensions about fitting in a Cessna 172. During the years that I had not been flying, I had gained a tremendous amount of weight and now was almost as big around as I was tall.

I went to the Aviation Medical Examiner and passed the physical examination. The flight school owner installed an extended length seatbelt in one of his 172s and I went flying.

The instructor that was chosen to check me out was from France and a little hard to understand, but we managed to communicate. I had butterflies in my stomach. I don't think he could tell.

We did the basics—steep turns, stalls, slow flight, forced landings, then a few landings. From a flying perspective, everything went well, but I was uncomfortable. I'm sure it was my size. I just didn't feel comfortable in the cockpit. The instructor was satisfied, though, and after two flights signed me off with a Biennial Flight Review. I was legal to pilot an airplane again. It had been almost fifteen years since I had flown previously.

The flight school had a Cessna 172XP, which I flew a few times. I liked the extra power it had, plus the fuel injected engine and variable pitch propeller.

Then I had the opportunity to fly copilot in a King Air B200 for a weekend. We started in Slidell, Louisiana, flew to Alexandria to pick up some passengers, then to Kansas City. From Kansas City, we flew to Shelbyville, Tennessee. Later that night we flew to Louisville, Kentucky. The next day we flew back to Alexandria, then took the plane back to Slidell. I didn't expect that I would be so rusty on IFR procedures, but I was. After that flight I bought a new copy of the Federal Aviation Regulations and the Airman's Information Manual and read them cover to cover. I vowed that if I was asked to fly in the King Air again, I would be on top of things.

My friend, Jim Halek, bought a Cessna 210 and I began flying it for him. The Cessna 210 could handle my size and it was a good traveling machine. Still, I wasn't really qualified to fly IFR without an instrument competency check. Then Jim bought a Cessna 421 and because of my previous Cessna 421 and 402 experience, I was qualified to be on the insurance policy provided I went to an approved recurrency training course.

We located a school in Champaign, Illinois, that offered both initial and recurrent training in the Cessna 421 and I flew the 210 up to Champaign to attend four days of training on the Cessna 421. The school included ten hours of simulator training in addition to fifteen hours of classroom work. Upon completion, I had a fresh Biennial Flight Review and an Instrument Competency Check, and was totally qualified to fly as pilot in command of the Cessna 421.

Life at Small Airports

It had long been a dream of mine to have an office at an airport. To look out on a runway and be able to watch the various airplanes coming and going would be, to my way of thinking, somewhat ideal. Be careful, they say, you might get what you ask for. Now that I was flying regularly again, I moved my office out to the airport.

Northwest Regional is home to over 550 airplanes. You can't tell that during a casual visit because none of the airplanes stay outside. They're all in hangars. There is a hodgepodge of hangars, too. They face all directions and come in many shapes and sizes.

I didn't know that there would be so many Stearmans in those hangars. I didn't know that on a pretty day, many of those hangars would send forth their occupants in the form of Wacos and Stinsons and Howards . . . RV-6s and Grumman Widgeons and Barons and Bonanzas and Cubs and Citabrias. Why, there must be 50 Bonanzas at Northwest Regional and on a pretty day, they all come out to play.

I didn't know that I would get up and go to the window every time I heard an engine run up, or that I would strain to see every plane that took off before it got out of site. I didn't know that on any given day, a B-25 or a Hawker Sea Fury, or a Learjet would make a high speed, low pass down our little runway. I didn't know that the Globe Swift and one of its RV-6 friends liked to take off together and fly formation at least a couple of times a week. I didn't know that one of those hangars had three Hueys in it and that one day I would get to hover one. Not fly it yet, because it's still not airworthy, but I did hover it for a few minutes.

I had no idea that so many people lived inside their hangars at that little airport. I didn't know that the pilot shop across the runway was the source of all information, both gossip and true and that it was worthy of a visit at least once or twice a day to find out what was going on. I didn't know, but should have, that flying and everything pertaining to it was so much a part of my DNA. Shame on me for letting those fifteen years slip by. I might have fifteen thousand hours by now. I may get there yet.

New York City

Date: 7/15/02
A/C Type: Cessna 421C
Registration #: N146TJ
Route of Flight: 52F – Republic Airport, Long Island

My first time to fly into New York City was in Jim Halek's newly-purchased Cessna 421C. Fortunately, I had a well-qualified copilot in Steve Petit who was a furloughed American Eagle pilot, formerly stationed at LaGuardia.

While buying charts for the trip a few days before the scheduled departure, I ran into a former New York Approach controller at the pilot shop. He advised me to stay under eighteen thousand and to file over the VORs coming into New York City and things would go pretty smoothly.

On the way up, we made a fuel stop at Lexington, Kentucky. I was at the controls when we arrived in New York's airspace and Steve was working the radios and doing the navigating. We got two routing changes from New York Center, with the final one taking us over Colt's Neck VOR in New Jersey at six thousand feet, then over the water past Staten Island and the Statue of Liberty and directly over the top of JFK airport. It was hazy, but still a very nice sight-seeing trip. It was much easier than I had anticipated.

When we left Long Island a couple of days later, we departed Republic Airport just at sunset and were again vectored over JFK. We got a beautiful view of Manhattan as the lights were coming on in the city.

Never fear flying to New York. Just stay out of the jet airways (unless you're in a jet), file in and out over the VORs (there are no SIDs or STARs applicable to light aircraft going to the outlying airports) and be prepared for routing changes.

ATC Confidence Builder—NOT!

Date: 8/14/02
A/C Type: Cessna 421C
Registration #: N146TJ
Route of Flight: 52F – Flying Cloud, Minneapolis

Steve Petit and I flew my friends Jim Halek and Franklin Santagate to Minneapolis where they were to speak at a pastor's conference. Flying Cloud Airport was right at the end of our comfortable fuel range, so when we arrived, we were anxious to get on the ground.

Runway 18 was the active at Flying Cloud, but it is only 2500 feet long. Steve was flying, I was in the right seat handling the radios and I informed the tower we preferred runway 27 Left, which was 5,000 feet long. "Enter downwind at your own risk," we were told. I gave Steve a puzzled look. I had never been told by a tower controller to enter the traffic pattern at my own risk. It was his job to sequence traffic in the pattern.

There was a Cherokee in the pattern for runway 18 and I saw him ahead of us as we turned downwind for 27. Suddenly, the TCAS announced, "Traffic" and another Cherokee showed up slightly below us and to our right on a converging path. Steve turned toward the airport to avoid him, while I asked the tower who he was and where he was going.

"I don't know," the tower replied. "He's talking to Approach."

Steve and I both found this a little disconcerting. Meanwhile, I was concerned about the Cherokee that was about to turn base for 18 just as we turned base for 27. "This isn't going to work," I told Steve.

"Six-Tango-Juliet, can you do S-turns on final, so I can get this Cherokee in?" the tower asked.

Steve started making S-turns, but he was already at 100 knots. We couldn't safely go any slower. I told the tower, "We're going as slow as we can go out here."

"Roger," he acknowledged, but still had us both heading for intersecting runways.

"I don't like this," I said to Steve. Tower should have the Cherokee go around. Finally, he did, just as we were about to abort our approach. It had not been handled very professionally.

A few minutes later, when we were taxiing to the ramp, I told the ground operator that we were holding short of runway 18, as we had been instructed to do. "Where are you on the airport?" she asked. Not a real confidence builder.

The next day when I called for our departure clearance, the ground controller advised us there was no flight plan for us in the computer, even though we had filed at least an hour earlier.

We began taxiing out while he worked on it. As I gave the controller the details of our flight plan again, he informed me that the computer would not take 52F, the identifier for Northwest Regional Airport, as it didn't recognize it. No problem. Just put AFW as our destination. Fort Worth Alliance airport is just seven miles away and it's a big airport, sure to be in the computers. Nope, wouldn't take that either. Try DFW, I suggested. That one wasn't recognized either.

We went ahead and took off under VFR conditions, but the controller assigned us 10,000 feet, which is an IFR altitude. Since the ground controller wasn't successful in getting us a clearance, I tried with the departure controller. We got handed off to one more departure controller and two Center controllers before we finally found somebody that was able to work around the computer issues and get us a clearance. Why the Minneapolis computers didn't know about the Texas airports is a mystery to me.

Partial Panel

Date: 1/13/03
A/C Type: Cessna 421C
Registration #: N146TJ
Route of Flight: 52F - OSU

We are always warned—by flight instructors, training publications, and articles in aviation magazines—that actual partial panel, when it does occur, won't be anything like what you've practiced in a training environment. Actual failures don't appear as gauges covered up by suction cups or sticky notes. Instead, they may happen gradually, so that an unsuspecting pilot doesn't recognize that the gyros have slowly wound down until he has followed them into a graveyard spiral. Or, they may occur as hard failures with or without warning flags.

In the old days when flying the Cessna 210 and the Turbo 206's, I practiced flying partial panel often, because those planes were notorious for vacuum pump failures. Fortunately, or maybe because of intervention from my guardian angel, I never experienced an actual failure in the clouds when flying those planes. Now I have.

We were going to Columbus, Ohio, in the Cessna 421. Jim, the airplane's owner, prefers to fly with two pilots, especially in the IFR environment, and I agree that it's a good idea, though the majority of my civilian IFR flying has been single pilot. On the morning of the flight, the other pilot called to say he couldn't make it. I assured Jim the flight would be fine with just one pilot. Even though it was IFR for departure with a ceiling around 600 feet AGL, the clouds topped out about 2,000 feet and we were heading for clear weather, which would begin approximately 100 miles north of our departure point.

Trouble started during the taxi and runup. The attitude indicator and horizontal situation indicator (HSI) flags weren't going off. They had been historically slow to come on line since Jim had acquired the plane six or seven months earlier, but always before, by the time the runup was done, the flags were off and the HSI would have slaved to the remote gyro that was located in the nose of the aircraft. Not this time, however.

During taxi, I checked, as I always do before an IFR flight, that the directional gyro turned as the airplane turned and that the attitude indicator dipped slightly when applying brakes. They both passed this test on this particular morning, but the heading on the directional gyro was not properly aligned and the warning flags were still on.

I obtained my departure clearance and a clearance void time via cell phone and taxied to the departure end of the runway, still believing that the instruments would soon come on line with sufficient warm-up time, just as they always had. The void time for the departure came and went and still the primary flight instruments weren't online. I called approach control back and got an extension on my time off, cycled the inverters back and forth a couple of times, and waited.

Then the flags went off—both of them. However the directional gyro was not properly aligned. I used the slew switch to align the directional gyro to the magnetic compass, then put the switch back in the slaved position. I then taxied onto the active runway and made sure the gyro followed me in both a slight right turn toward the end of the runway and a ninety degree turn to the left to line up with the runway centerline. As a last minute precaution, I reached across to the copilot's side and aligned the directional gyro there with the runway heading and set that altimeter to the field elevation.

Satisfied that there were no warning flags and that the gyros were properly aligned, I ran the engines up until the airplane was straining at the brakes, released the brakes and off we went.

At approximately 1200 feet MSL (600 feet AGL) we went into the clouds. A minute or so later, I started pulling the throttles and prop levers back to their climb settings and contacted approach control. My clearance had been to fly heading 030 after departure. Since we had departed runway 17, I was already starting a left turn to the assigned heading, when the controller came back with instructions to fly heading 270. I initiated the right turn, still climbing and still adjusting the prop levers. We heard a sound, Jim and I both did, like racing engines and he said, "Something's wrong." I felt it, too.

Even though I'd initiated a right-hand turn, the directional gyro was still on the runway heading and the attitude indicator was showing a 10 degree nose up. That couldn't be right! Not with the control inputs I had made and with the sounds and sensations that both Jim and I were hearing and feeling.

Quickly, I looked over at the copilot's instruments and saw the directional gyro spinning to the right and nothing but blue on the attitude indicator. All of my unusual attitude recovery training immediately came into focus as I first leveled the wings, then began slowly bringing the nose up. As I was recovering from what was obviously a tight right-hand sprial, we bottomed out of the clouds, which meant that we had lost several hundred feet in just a few seconds. The wings were by that time level and the nose was approaching level, but just the fact that we saw the ground so close in front of us made it even more scary.

Now in control, I began climbing again, this time using the copilot's instruments for reference. Approach was giving me more instructions, which wouldn't work. The controller cleared me to "fly heading 270 until intercepting the Maverick 360 degree radial, maintain 3,000." Since the Maverick VOR was east of our location, that clearance would never get us there. Besides, the clearance was supposed to be to one of the outbound intersections normally used when departing the DFW area to the north from an airport not included in the Standard Instrument Departures (SIDs). I advised the controller that I needed to go to the

BLECO intersection and that his heading would never get me there. He apologized and assigned me a 360 heading temporarily.

Passing through 2,000 MSL, we flew into bright sunshine. About that time, the directional gyro on the HIS directly in front of me "unfroze" and spun around to our correct heading. It behaved properly for the rest of our trip.

We made it to Ohio and back safely, in spite of the fact that we encountered an unexpected snowstorm while on the ground at Columbus. The instruments worked fine for the rest of the trip, but I was highly suspect of the remote gyro in the nose of the aircraft. The only other incident of note was the fact that a door warning light came on fifteen or twenty minutes after our departure from Columbus. It had to have been my fault, as I was the last one to board the airplane and the one that had closed the door prior to departure.

We were in the clouds, so I was on the gauges. Jim and the other passenger in back tried to close the door by pulling on it and opening and closing the latching handle, but were unsuccessful. I advised Dayton Approach, whose area we were then passing through, that I needed to make a precautionary landing to close a door. He offered me a choice of airports, the big one at Dayton, or the Wright Brothers Memorial Airport. Since the latter was closer, I chose it and immediately got a vector to the airport and a descent clearance.

The ceiling at the airport was 1500 broken and the visibility was three miles or better in haze, so the controller was counting on me seeing the airport and being able to execute a visual approach. That's exactly what happened, though the airport was actually a little behind me when I saw it. I turned for a left downwind to land north on the north-south runway, slowed the airplane and lowered the gear. No green light for the nose gear. Just what I needed. Jim was in the back, holding the door closed with his hands and the gear wouldn't go down. I raised it and lowered it again. Still no green light for the nose gear. I did not want to have to blow the gear down with the nitrogen bottle and then have to have maintenance done before we could depart again. I raised and lowered the gear one more time and this time I got three green lights, just as it was time to turn base. I cancelled the IFR clearance and turned final. I never did get a chance to switch to the advisory frequency or even look it up. I noticed a Citation waiting to depart, but I was now on short final with three landing lights and two recognition lights on, so I trusted that the Citation pilot could see me.

Back in Texas, I discussed the occurrences with my friend Glen Hyde, a former airline pilot, accomplished warbird pilot, an Airframe and Powerplant (A & P) mechanic, and owner of Northwest Regional Airport. Glen's diagnosis was that either the remote gyro was shot or that the amplifier that sent its signals to the display in the cockpit was bad. When I told him about the door incident, he asked if I had thought to turn off the pressurization when the guys in back were attempting to shut the door while we were airborne. Such a simple thing and it made all the sense in the world since part of the pressurization system's job is to inflate a rubber seal around the door to keep air from escaping from the cabin. I hadn't thought of that, but will the next time, if there ever is a next time. Hopefully, there won't be.

On the first trip to the avionics shop, they said they didn't find anything wrong. I was highly suspect and so was Jim. I remembered that Field Tech Avionics in Fort Worth had always given me great service back when I was managing a fleet of airplanes based at Meacham, so we gave them a call.

Field Tech's treatment of the problem was very professional, but even they didn't catch the entire problem the first time around. They bench checked the remote gyro and found that it was only spinning up to 800 rpm. The minimum specification called for it to spin at 1200 rpm within three minutes of applying power. They installed a rebuilt gyro, and checked a few connections, and still got the warning flags.

The rebuilt gyro was doing its job, so there had to be another problem. They pulled the AN112, which is the display unit that goes in the instrument panel that contains both the attitude indicator and the HSI. They hooked it up to a test unit on the bench and began sending signals to it. When mimicking a left turn, everything worked great, but when they applied the signal to indicate a right-hand turn, the instrument froze for a few seconds, after which it broke free and properly indicated the turn.

That was exactly what I had experienced in the air while flying IFR. The interesting thing about this failure was that with the proper amount of power and signal applied, the warning flags were off, yet the instrument didn't properly indicate a turn.

Field Tech didn't have a rebuilt unit in stock, but one of the technicians offered to come in on the next day, which was Saturday and rebuild the unit. I was impressed with Field Tech's commitment to get us flying as soon as possible.

On Monday, we flew the airplane to Port Arthur, Texas, about an hour and fifteen minutes away. Everything was fine on the flight down, but during the flight back the warning flags for both the attitude indicator and the HSI were on during the entire flight, though they seemed to be operating normally.

The next morning, I took the plane back to Field Tech and waited around while they checked things out. What they found this time was that a particular component of the AN112 unit's power supply, passed all static tests, but failed when a vibration was applied to the unit, such as in flight. That component was replaced and the problem solved.

I was grateful that my instrument experience and training paid off when the instruments failed in flight so that I automatically knew to turn to the backup instruments on the other side of the cockpit for reference. Why did both the attitude indicator and HSI fail at the same time? Because the power supply within the AN112 is common to both indicators.

I learned several things from the experience. N146TJ has a Collins Flight Director and Autopilot system that is at least twenty years old and there was no documentation with it when we got the airplane. Nor could I find any manuals for it on the Internet. Consequently, I didn't know what the parameters were for determining safe operation. I learned from the avionics shop that if the gyro isn't up to speed within three minutes, something is wrong with it. I also know that using the slew switch to line up the gyro isn't the way to do it. If it doesn't slave itself, then it's not operating properly. Knowing what I know now, I would have never taken off from Northwest Regional in IFR conditions with the slaved gyro system behaving as it had. That's what experience is all about.

Engine Failure - This Time I Had a Spare

Date: 6/7/03
A/C Type: Cessna 421C
Registration #: N146TJ
Route of Flight: 52F – Tyler

The first reaction was one of disbelief. "Brad, I do believe the left engine just quit." On takeoff you expect an engine might quit. You're constantly thinking, "If it quits now, here's what I'm going to do . . . ." Then after flying speed is reached and there's not enough runway left to stop, "here's what I'm going to do . . . ." Again, after the gear is up, but before the power is reduced, "if it quits now, here's what I'll do . . . ." But not at cruise. At cruise, you're occupied with charts and fixes and settings and radios. You're not thinking constantly about an engine failure.

Besides, at cruise engines give you warning. They sputter or cough or surge to let you know that you forgot to switch the fuel selector or there is ice blocking the air inlet, or the fuel pump is going south. They don't just quit.

This one did. We were at 13,000 feet, just intercepting the airway direct to GGG VOR, watching the course lines change on the navigators and making final adjustments to the autopilot, when the sound abruptly became quieter and there was a momentary yaw to the left. Not a big yaw, just a slight yaw which the autopilot quickly corrected.

The disbelief reaction was a short one, probably less than a second or two, then the training kicked in. Verify the power loss, which was not hard in this case because the left propeller was slowly winding to a halt and the tachometer needle with the big L on it was winding down. Quick, feather the prop while you still have oil pressure, then move the mixture and throttles to idle. That I quickly did, then turned off the fuel pump for the left engine.

The engine was secured, but was it the correct one? Yes, the right engine was still moving us along quite nicely. We were maintaining altitude, though the airspeed had dropped to around 160 knots.

It was a nice time to have a copilot. "Brad, tell ATC we've lost an engine, then find us a place to land." Three high-priced navigators on the panel, each with a *nearest* feature and still the fastest way to find a suitable landing place was to simply ask. "Center, this is Golden Eagle one-four-six-tango-juliet. We've just lost our left engine and need to proceed to the nearest airport with facilities."

"One-four-six-tango-juliet, say again your equipment failure."

"We've lost our left engine. We'll we haven't actually *lost* it. It's still there on the wing; it's just not running."

"Roger, Tyler is 34 miles southeast."

"We'll go there, then."

"One-four-six-tango-juliet is cleared direct Tyler, maintain six thousand. Will you be needing any assistance?"

"Out of one-one thousand for six thousand, direct Tyler, no it looks like everything is under control here. Thanks."

And it was. I continued to clean up the airplane by rolling in enough rudder trim to keep the ball centered, plus five degrees of aileron trim toward the good engine, and along she motored. I started down. Later, when safely on the ground, I thought about how I should have tried a climb, just so we would know what a climb was like on one engine, but at the time, my only interest was in making a safe landing.

Before we could level at six, we were handed off to Tyler Approach and cleared on down to three thousand. We then saw the runway and were cleared for a visual approach. It was a perfect straight in approach and though I delayed the flaps and gear, I kept having to decrease power on the remaining engine to get the airplane to go down.

With the runway made, I eased off all of the power and the nose started swinging around to the right, compliments of all of the rudder trim I had cranked in. It was an easy correction, and then we were down. We taxied off the runway at the end, but I gave up on trying to get the plane to the ramp. Too much asymmetrical thrust to taxi straight ahead, so I called for a tug.

The diagnosis when the engine was torn down was pretty bad. Basically, the engine self-destructed from the inside out as a result of restricted oil flow to the main bearing. That happened because of a previous maintenance issue that had caused the airplane to lose much of its oil through the oil cooler connection on a previous flight. The leak had been repaired and the remaining oil examined for metal particles. None were found and the airplane had been signed off and returned to service.

Naturally there were questions. Did you have oil pressure? Yes, we did. Maybe not in the minute or two immediately prior to the seizure, but every time we looked the oil pressure was in the green. Nobody stares at the engine gauges every minute of every flight. We checked the gauges on run-up, we checked them on takeoff, we checked them as we climbed, and we checked them upon level-off. Each time we looked the instruments were in the green. Maybe that changed just before the engine failed, maybe it didn't. Without a recording engine monitor, we'll never know. Maybe airplanes should have a big red oil pressure low light right in front of the pilot like cars have to get the pilot's attention. Naturally, you want to save an engine if you can by shutting it down at the first sign of trouble as long as you have plenty of altitude and airspeed and a spare engine. But in this case, we had no warning.

For Now There is Flight Simulator

As this book goes to print, I've come to another chapter in my flying career. I certainly hope it's not the last one. The flying opportunities afforded me over the past three years have now passed into the hands of younger men who have chosen careers in aviation. I'm all for them. May their wings stay free of ice, may their weather equipment always work, and may their destinations all hold promise.

Microsoft's new Flight Simulator 2004 has enough aircraft and enough challenges to help me keep my instrument skill sharp for the present. I'm working to maintain and even improve my health and believing for more flying opportunities in the future. Why just this week I saw an ad in Dallas for a medical evacuation pilot with Cessna 421 experience. Who knows?

My real dream is to use my aviation skills to help spread the good news that Jesus Christ, God's Son, came to earth to free us from the curse of sin and death. I've had opportunities during the last three years to fly pastors and evangelists to meetings, helping them to meet demanding schedules that they could not have met using other forms of transportation. To me that is a very rewarding way to utilize the skills and experience I've had the opportunity to develop. It is my prayer that such opportunities will arise in the future.

Glossary

Pilots are the worst about using jargon. It's language that other pilots understand, but to the non-aviator there may be confusion. I tried to define most of the acronyms and jargon terms within the text, but just in case I missed a few, here's a glossary of terms:

AGL

Above Ground Level - a measure of altitude (see also MSL).

Antitorque

As used in this book, the term describes a procedure in which the helicopter pilot uses the throttle and slipstream to maintain the longitudinal stability of a helicopter fuselage after experiencing a failure of the tail rotor system, whose job it is to counteract the effects of torque from the engine and transmission.

AO

Area of Operations (military term).

Autorotation

When a helicopter's engine fails, the transmission that connects the engine to the rotor system will allow the rotor to turn freely. Naturally, without power, the helicopter will begin descending. During the descent, the airflow over the rotor blades will keep the rotor system turning so that there is energy left in the rotor system that can be used to cushion the landing. This is called "autorotation" and is the method by which a helicopter pilot can land safely after an engine failure. Helicopter pilots in training practice this manuever frequently and under a variety of conditions.

C & C

Command & Control.

CAVU

Clear and Visibility Unlimited.

Charlie

Slang term for Viet Cong during the Vietnam War.

Contact Approach

When on an instrument flight plan, if the pilot can see the runway he can request a contact approach which allows him to land without following a published instrument procedure.

Crab

Short for "crab angle" which means that the nose of the airplane is turned into the direction of the wind to keep the aircraft flying along a particular course on the ground, i.e., to keep the wind from blowing the airplane off course.

DEROSE

Date of Expected Return from Overseas—when you got to leave Vietnam.

FAC

Forward Air Controller. These guys directed airstrikes and artillery during combat.

First-Up, Second-Up

This is a duty crew designation. First-Up would be the crew to take the first or primary mission, while Second-Up would be the backup crew.

IP

Instructor Pilot.

Litter

The stretchers used in military helicopters and ambulances to transport patients.

LORAN

Long Range Navigation. This is a ground-based navigation system that allows direct flight, similar to GPS, which is based on satellite navigation.

Mattel Messerschmidt

The TH-55 Army training helicopter had some parts made by the toy company Mattel, hence it was sometimes called the Mattel Messerschmidt.

MSL

Mean Sea Level. The altitude above sea level (See also AGL).

OBS

Omni-bearing selector. A cockpit navigation instrument used to navigate along airways consisting of radials eminating from VHF radio transmitters.

Peter Pilot

A slang term used to describe the First Pilot (copilot) in an Army helicopter crew.

Pirep

Pilot report on weather or other flying conditions.

Rime

Ice that is made up of many crystals forming a rough surface.

RMI

Radio Magnetic Indicator. An instrument used for navigating along VHF airways.

Sortie

A single mission.

Sitrep

Situation Report.

SAM

Surface-to-Air Missile.

Sandy

Nickname for the Douglas A1E Skyraider, a single-engine, propeller-driven airplane used for close air support.

VOR

VHF Omni Range. A navigational radio beacon used to designate airways.

www.ingramcontent.com/pod-product-compliance
Lightning Source LLC
Chambersburg PA
CBHW032057050726
47590CB00001B/308